ALLE ZEIT WACH
1842

Ekkehart Schlicht

Isolation and Aggregation in Economics

Springer-Verlag
Berlin Heidelberg New York Tokyo

Prof. Dr. Ekkehart Schlicht
Institut für Volkswirtschaftslehre,
Technische Hochschule Darmstadt, Schloß
D-6100 Darmstadt

ISBN-13:978-3-642-70300-3 e-ISBN-13:978-3-642-70298-3
DOI: 10.1007/978-3-642-70298-3

Library of Congress Cataloging-in-Publication Data
Schlicht, Ekkehart, Isolation and aggregation in economics. Bibliography: p. Includes indexes.
1. Macroeconomics. I. Title.
HB172.5.S35 1985 339 85-12638
ISBN-13:978-3-642-70300-3 (U.S.)

Softcover reprint of the hardcover 1st edition 1985

2142/3140-543210

For my family

Preface

In order to solve a given problem, economic analysis is compelled to concentrate on the interaction of selected factors while disregarding a multitude of other influences. This book offers a discussion of certain central premises involved here and draws some analytical consequences. The argument is focused on process analysis, i.e., on the analysis of economic processes within a given institutional setting, although certain corollaries for institutional analysis are patent.

Many colleagues and students have helped me, for many years, to develop the views presented here, and it seems impossible to trace individual influences. Thus I can only express my indebtedness in a macro sense.

I wish to thank the Westdeutscher Verlag for its kind permission to use material from my *Grundlagen der ökonomischen Analyse.* The results of Chap. 4 were presented at the Econometric Society European Meeting in Pisa, 1983. Dr. W. A. Müller from Springer-Verlag has encouraged me to write this book and has been helpful in many ways.

Darmstadt, March 1985 Ekkehart Schlicht

Table of Contents

1. The Setting of the Argument

> **The laws of economics are to be compared to the law of the tides, rather than with the simple and exact law of gravitation.**
>
> **Alfred Marshall**

1.1 Two Characteristics of Economic Analysis

Economic phenomena are the outcomes of a plethora of factors, and economic analysis, unable to tackle them all, is compelled to select those factors which seem to be the most important, and to consider all other influences as data of the analysis. But these data are only provisional since they are wandering themselves. One characteristic of economic analysis is, then, that it is built on a moving foundation.

Furthermore, economics is not concerned with the idiosyncrasies of particular cases, but looks for general rules linking typical incidents. Usually these rules cannot be distilled stringently from the multitude of individual actions, and economics is bound, hence, to start from assumptions on the behaviour of aggregates – or "representative" agents – which are linked to individual actions just vaguely. This way of proceeding is most apparent, and most important, in the case of macroeconomics, where we assume aggregate behaviour from the outset, but it is also of relevance for microeconomics.

The following discussion sets out to probe into these problems: the provisional nature of economic data and the nature of aggregative assumptions. Since we are going to tackle these problems by isolation and aggregation, these considerations set the themes for the book.

1.2 The Provisional Nature of Economic Data

1.2.1 Data and Variables

We explain economic phenomena by relating them to their causes. Price formation in a particular market is, for instance, explained by the interaction of supply and demand. The conditions of supply and demand are the *data* of this problem which are presupposed in the explanation of the *variable,* which is the price in our example. Quite generally those conditions which are presupposed in the analysis are the data, and those phenomena we seek to explain are the variables of the problem [1].

1.2.2 Provisional Data

An economic datum is not, however, absolutely fixed like a natural constant. Rather it is to be viewed as being determined by other causes: The supply behaviour of firms is influenced by their strategies and the technological conditions under which they operate, and the demand of households is influenced by their incomes and preferences.

Again, these are not ultimate data since preferences, strategies and technologies are subject to various influences and might change over time. Hence economic data are typically only of a provisional nature. This holds true quite generally since economic processes are built on psychological factors and are embedded in and interact with sociological processes. Hence psychological and sociological circumstances are to be considered as data, and we know that these change over time: They have a dynamic of their own and are themselves influenced by the economic processes.

But even if we aim to incorporate all relevant psychological and sociological processes within the analysis, we cannot hope to gain a firm stand. We will be unable to predict the development of all those influences subject to technical change, since we cannot predict still unknown future inventions with accuracy [2]. Hence those data describing technology are provisional by necessity, and the provisional nature of economic data cannot be avoided in this respect, at least. Since technological change is of paramount importance in economics, we have to face this difficulty, and we have to face it in practice also with respect to sociological and psychological data since there is no hope, at present, for a useful all-embracing theory of social processes.

The provisional nature of our data leads to a particular problem: What is to be considered as variable, and what is to be considered as

1 Variables can be quantities, but also qualities: Private property can be viewed as one particular form of ownership, and we can try to explain why this form has come to be. Here the form of ownership is the variable of the problem.

2 At best we can make some broad guesses about future developments: Vague aggregative predictions might be possible here, but no detailed predictions excluding changes and deviations.

fixed if everything is floating? What are the requirements which are to be met by those factors which we take as data of our analysis? Let us consider this question.

1.2.3 The Ceteris Paribus Clause

Economic data are not ultimate data, like the speed of light in physics. Rather they are provisional in nature. This is expressed by means of the *ceteris paribus clause.* All factors not explicitly considered as variables are assumed to be fixed within an argument. This clause is used, explicitly or implicitly, throughout economics.

The ceteris paribus clause is particularly restrictive in those cases where only a narrowly limited issue is analysed, such as price formation in a single market. Effects on other markets, and possible repercussions, are excluded. Price changes in one market might lead to price changes in other markets, however, and these work back on the market under consideration. The assumption that all other prices are given is certainly wrong here. But the same reservation applies also to more comprehensive theories, since these have to presuppose data, too, and repercussions of the processes under study on the data of the analysis cannot be excluded, irrespective of how we try to delimit the problem [3]. Economics is bound to perform *partial analysis* rather than *total analysis:* It considers phenomena in an economic system which is only a part of the wider complex and interdependent social system, and fixes its demarcations by means of the ceteris paribus clause.

1.2.4 A Difficulty in Applications

An economic theory can offer only an analysis of a partial issue and proceeds from provisional data which are generated by means of the ceteris paribus clause. These provisional data will change over time, however, contrary to what was postulated. This leads to a particular difficulty in applications: Wrong predictions can always be attributed to changes in boundary conditions while maintaining that the theory would have given a correct prediction if the ceteris paribus clause had not been violated.

---

3 Walras seems to have overlooked this point. He has argued against the study of isolated markets with the above argument, but failed to see that this argument can be turned against general equilibrium theory as well (see Walras, appendix II).

If it is predicted, for instance, that certain policy measures will boost the economy, and the expected economic recovery does not actually occur, it is always possible to name certain "special and unforeseeable" developments which can be made responsible for the failure of the policy measure without invalidating the theory.

Even if a prediction turns out to be correct it can always be argued that the predicted phenomenon has occurred for other reasons than those given in theory. Jevons (p. 88) gives the following example.

> Entirely to prove the good effects of Free Trade in England, for example, we ought to have the nation unaltered in every circumstance except the abolition of burdens and restrictions on trade. But it is obvious that while Free Trade was being introduced into England, many other causes of prosperity were also coming into action – the progress of invention, the construction of railways, the profuse consumption of coal, the extension of colonies, etc., etc. Although, then, the beneficial results of Free Trade are great and unquestionable, they could hardly be proved *a posteriori*[4].

The complexity and diversity of the factors fixed under the ceteris paribus clause leads, hence, to our inability to control them. We are not even able to name them in detail; to describe any state of the world completely seems to go beyond our powers.

1.2.5 Economics as a Deductive Science

Jevons draws the following conclusion from his example.

> Although, then, the beneficent results of Free Trade are great and unquestionable, they could hardly be proved to exist *a posteriori:* they are to be believed because deductive reasoning from almost certain truth leads us confidently to expect such results, and there is nothing in experience which in the least conflicts with our expectations (Jevons, p. 88).

4 This problem is a matter of principle even if we are able, nowadays, to set up an econometric model separating these effects. This is the well-known identification problem in econometrics which arises always because there exists in all cases an infinite number of theoretical models compatible with a given set of observations. In order to solve this identification problem, we have to restrict the class of admissible models on a priori grounds (by using the principle of parsimony, for instance) and to develop a priori criteria for the goodness of fit such that the observations select one and only one of the admissible models. By the way Jevons seems to presuppose in the above example that the various influences mentioned combine additively. In the words of John Stuart Mill (p. 59) economic forces are viewed as combining *mechanically* rather than *chemically* here, i.e. they are not assumed to produce qualitatively novel phenomena by their interaction. As pointed out by Marshall (p. 637), however, this might be misleading.

If economics is viewed in this spirit as a deductive science drawing logical implications from certain premises, it cannot lead to false conclusions, but this shifts the problem to the assumptions we start with: How can we convince ourselves that these assumptions are reasonably realistic, if this cannot be done with reference to their implications, as Jevons has argued, and if we know that the data we start with are of provisional nature only?[5] Critical introspection might give some hints as to rules of human conduct, no doubt, but it will highlight, too, the volatile nature of our motives and our conduct, and the various influences interfering here. To gain a firm starting point (if there is any) would require to go into psychology[6].

1.2.6 The Homo Oeconomicus

One way out is to confine economics to the *economic aspect* of behaviour. This is the view of Carl Menger. According to this view, economics is concerned with the course of events under the hypothetical assumption that man acts as a *"homo oeconomicus"*, guided solely by economic motives – the striving for the maximization of utility and profit. The existence of other motives is not denied, neither is it excluded that their effects might override the economic factors, but their workings are not seen as a theme which exact economics has to consider; this theme is relegated to the other social sciences[7]. The relevance of this view is defended by asserting that economic motives

5 I do not consider Friedman's view here that it does not matter whether the assumptions made are realistic or not as long the predictions are correct – I am not interested in theories leading to right predictions for the wrong reasons (cf. Friedman).

6 Some economists think that we are entitled to postulate stable time-invariant preferences of individuals, if these preferences are suitably formulated; this artifice might avoid psychology to a large extent (see Stigler/Becker). The idea is that commodities are thought of as producing satisfaction not directly but indirectly through a household technology which transforms these commodities into variables entering the utility function. Learning effects can be described by this approach, for instance, if the consumption of music produces "music appreciation" by increasing the faculty for music appreciation simultaneously with its consumption. In this way all instabilities can be attributed to changes in household technology rather than to changing preferences. Although this view is sometimes very illuminating, it shifts the psychological problem to household technology and replaces the phrase "tastes remaining the same" by "household technology remaining the same", i.e. it cannot do away with the ceteris paribus clause.

7 Menger p. 59. Menger distinguishes between "exact" or "theoretical" and "realistic empirical" economics. The latter is only able to establish historical regularities without proving their necessity.

are *dominant* in propelling the economic process, and it is hence important to study their workings in isolation[8]. This claim for relevance can only be maintained, however, if we can explain why it is admissible to fix all "non-economic" factors under a ceteris paribus clause even if they are only of minor importance, and still obtain meaningful results.

1.2.7 The Isolation Principle

If an economic theory is intended to elucidate real issues, we have to use ceteris paribus clauses which are sufficiently realistic. This condition will never be exactly fulfilled. Hence we cannot hope to devise exact theories. Nevertheless our theories can give an appropriate picture of the process under study if the data of our analysis are reasonably stable in comparison to the speed of that process.

We are entitled, for instance, to explain the level of a certain price by the intersection of supply and demand schedules at each instant of time if price adjustment is very fast and the changes in the configuration of supply and demand are very sluggish. Here the ceteris paribus clause "supply and demand conditions remaining the same" will lead to an equilibrium price which approximates, at any instant, the true price sufficiently well. This approximation will not be possible, however, if price adjustment is slow and supply and demand conditions are changing rapidly. The ceteris paribus clause "supply and demand conditions remaining the same" will not be admissible in that case.

More generally it is to be required that those factors fixed under a ceteris paribus clause be sufficiently stable with regard to the process under study[9]. This is the *isolation principle.* Chapter 2 ("On Isolation") is devoted to its elaboration.

1.2.8 The Moving Equilibrium Method

A basic premise of economic analysis is that a result derived under a ceteris paribus clause will remain approximately valid even if the factors fixed under that clause are actually changing slightly. This premise is, however, not always true, but rather only under certain conditions, which our theories must meet.

8 Menger p. 79.
9 Stable in the sense of sufficiently invariant over time.

Chapter 3 ("The Moving Equilibrium Method") is devoted to a discussion of this proposition in the context of equilibrium analysis: Moving data lead to moving equilibria – hence the name of the method – and the question is whether actual movements can be approximated sufficiently well by moving equilibria.

1.2.9 The Nature of Economic Theories

Since economics has to build upon provisional data which are generated by the ceteris paribus clause and satisfy the isolation principle, it can only hope to explain phenomena by factors which are changing themselves, but changing more slowly. Keynes put it as follows.

> Economics is a science of thinking in terms of models joined to the art of choosing models which are relevant to the contemporary world. It is compelled to be this, because, unlike the typical natural science, the material to which it is applied is, in too many respects, not homogeneous through time. The object of a model is to segregate the semi-permanent or relatively constant factors from those which are transitory or fluctuating so as to develop a logical way of thinking about the latter, and of understanding the time sequences to which they give rise in particular cases. [10]

1.2.10 Consequences for Econometrics

This view of economic theory as forcibly treating fluctuating objects has implications with regard to empirical applications: Economics presupposes that the economic variables under consideration are governed by laws which are changing over time. The data are the coefficients entering the laws. Since the data are not explained themselves – else

---

10 Keynes (3) p. 296ff. Incidentally, this view explains the striving for generality in economics which puzzles philosophers of science. Hamminga (p. 8) writes, for example: "If a theorem holds in many worlds that can be expressed in the economist's language the probability of a theorem to be true in our real world is high, even without considering at all what our real world exactly is like", and he finds this view somewhat funny although it underlies, according to him, the economists' notion of scientific progress. The underlying – and sound – view of the economists is, of course, that we cannot presuppose a definite real world since we cannot know it, and since it is everchanging. Hence only theorems which remain valid in a changing world are of interest. Hayek stresses the same by saying that economic theories are bound to be "algebraic" for that reason, and Keynes writes likewise: "To convert a model in a quantitative formula is to destroy its usefulness as an instrument of thought ... To do so would make it useless as a model" (see Hayek ( ) p. 338 and Keynes (3) pp. 299, 296).

they would be endogeneous variables – they must be permitted to change erratically. According to the isolation principle they are permitted to change only slowly. Hence this slow change of the coefficients is the basic premise econometrics has to start from. Chapter 4 ("Econometric Implications") discusses a method for estimating economic relations starting from that assumption.

1.3 Intricacies in Macroeconomic Analysis

1.3.1 Economic Aggregates and Macroeconomic Problems

In macroeconomics we are interested in the determinants of broad economic aggregates: What determines the price level, the level of unemployment, the growth rate, the balance of payments? We might even pose similar questions on a slightly more disaggregated level: What determines the price level of a certain group of products, the level of unemployment of a certain group of individuals, the growth rate of a particular industry, or its contribution to foreign trade? But we are not so much interested in the particular prices, in the employment situation of a particular individual, or in the foreign trade balance of an individual firm.

There are two types of economic aggregates: Aggregate quantities and aggregate agents[11]. Aggregate quantities are those numbers we typically deal with in macroeconomics: gross national product, labor productivity, the price level, the rate of unemployment, and so forth. These are broad indices summing up various microeconomic quantities. Aggregate agents are obtained by grouping together microeconomic agents and by conceiving these groups as units acting and reacting in certain ways. Typical aggregate agents are industries, sections of the work force, sectors of the economy, and the government.

Macroeconomic problems, then, relate to the determinants of the behaviour of economic aggregates: What determines the price level?; what determines the behaviour of an industry?; etc.

---

11 One might, perhaps, also conceive of something like "aggregate institutions", but since we will not consider institutional analysis at all, this discussion is omitted for the sake of simplicity.

1.3.2 The Microeconomic Approach to Macroeconomic Problems

One can try to answer macroeconomic questions by considering explicitly the microeconomic quantities and agents and looking for the corresponding macroeconomic implications. If macroeconomic problems can be tackled successfully in such a way, this is very satisfactory; but very often they cannot.

The reason is, basically, the following: The most general microeconomic models are so all-inclusive that they can explain virtually all macroeconomic outcomes[12]. In order to obtain definite conclusions with regard to aggregate consequences, we have to impose additional microeconomic restrictions on them, but it is very difficult to formulate restrictions which can be imposed on the behaviour of all microeconomic agents, and it is practically impossible to obtain all the necessary data to verify them, let alone to describe the behaviour of all microeconomic agents exactly[13].

1.3.3 The Macroeconomic Approach to Macroeconomic Problems

To avoid the difficulties of using microeconomic approaches to macroeconomic problems, macroeconomic theory looks at the behaviour of aggregate agents and aggregate quantities directly. This approach has the definite advantages that the problems that interest us are put into focus from the beginning, and economic aggregates can be observed much more easily, since official statistics are available. It has, however, the definite disadvantage that we do not know how the behaviour of the macroeconomic agents comes about since it is explicitly left open how microeconomic activity leads to exactly those macro relations we start with.

---

12 As Sonnenschein and Debreu have shown, the general equilibrium model implies essentially no restrictions on the shape of the excess demand functions and hence on possible equilibria. Lexis has argued alike as long ago as 1895. The same holds true, *a fortiori,* for the so-called "Keynesian" generalizations of the general equilibrium model (Benassy) since they include general equilibrium as a special case.

13 There is one promising attempt to derive properties of market demand functions by putting restrictions on income distribution, rather than on individual demand functions (Hildenbrand). Note, however, that the strategy here is to describe a class of microeconomic models leading to a presupposed macro outcome, and postulating the macro outcome directly involves the smallest number of microeconomic restrictions. Providing some possible microeconomic underpinnings raises, however, the plausibility of the direct macro assumption.

With regard to many important questions, macroeconomic analysis is simply unavoidabe since no microeconomic theories are available which could be used: Business cycle theory, foreign trade theory, unemployment theory, the theory of distribution, or monetary theory proceed in aggregate terms from the outset, for instance, and we have only macroeconomic tools.

1.3.4 The Hermeneutic Aggregation Problem

If we do macroeconomic analysis – that is, aggregative analysis without explicit microeconomic foundations – several problems arise. The first is: How can we expect stable macroeconomic relations to exist? Are not those macroeconomic postulates we start with merely fictions as long as no explicit microeconomic foundation is provided[14]? Or is it less demanding to postulate the existence of stable macroeconomic relations than to postulate stable microeconomic relations? Further, how should we envisage the connexion between microeconomic and macroeconomic laws in principle, and what is the nature of those macroeconomic relations we work with? All this constitutes what might be termed the *hermeneutic aggregation problem.* It is to be distinguished from the *technical aggregation problem* which consists in describing a *known* microeconomic system in aggregate terms.

Chapter 5 deals with the aggregation problem from the hermeneutic perspective. It argues that macroeconomic analysis proceeds by employing certain ceteris paribus clauses, and that aggregative relations can be described by a generalization of the moving equilibrium method. Furthermore it is argued that macroeconomic relations are generally more stable than microeconomic relations.

1.3.5 The Typical Agent

But how can one generate hypotheses concerning the behaviour of aggregate agents? This is usually done by conceiving typical microeconomic agents – a typical firm, a typical household – and by assuming that all microeconomic agents are typical agents. It is usually rather easy to deduce the behaviour of aggregate agents from these premises, and the resulting aggregate relations are the relations we start with in

14 see Menger, p. 87.

macroeconomics. This is quite reasonable if we can safely assume that the behaviour of the typical agents dominates the behaviour of the aggregates, and the various microeconomic idiosyncrasies either do not matter too much macroeconomically, or else cancel each other. Furthermore, the construction makes sure that the macroeconomic relations we start with involve no logical contradiction, since there is at least one microeconomic model – namely that involving only typical agents – which is compatible with our macroeconomic hypotheses. It is of course never maintained that all microeconomic agents actually *are* typical. The notion of a typical microeconomic agent is simply a theoretical device for deriving plausible macroeconomic assumptions.

1.3.6 The Representative Agent

If we have obtained hypotheses on the behaviour of aggregate agents in this manner, we might ask ourselves whether we can look at these agents as following certain maxims of behaviour and acting under certain constraints such that their behaviour can be explained by this construction. Is the behaviour of an industry such that we can envisage it as being the behaviour of one single firm striving for profit maximization (or something else), or can the behaviour of the household sector be conceived of as the behaviour of one single household striving for utility maximization (or something else)? If this can be done, we call the agent a *representative* agent, and such a construction might be very helpful in elucidating macroeconomic issues.

1.3.7 Micro and Macro Behaviour Might Differ

If all households are identical, and no interaction occurs among them, the behaviour of the typical household will appropriately describe the behaviour of the household sector. In such a case the typical household can be taken to be the representative household.

But in many cases it is quite misleading to identify the typical and the representative agent. An industry might grow although all firms shrink in size, and the consumption habits of individual households might be influenced by the consumption of other households, but no reference group effects are possible for the representative household, since it stands alone. Examples such as these give only the simplest cases where the behaviour of the typical agent deviates from that of the corresponding representative agent, but they illustrate already that micro and macro behaviour might differ. Sometimes this difference might even be more drastic. We might envisage representative agents for instance having no counterpart in any typical microeconomic agent at all. The discussion of the hermeneutic aggregation problem in Chap. 5 is intended to contribute to a proper understanding of these somewhat complex micro-macro transformations [15].

1.3.8 Microeconomics and Aggregation

The aggregation problem is most conspicuously present in macroeconomics and will be discussed in that respect in this book. It seems to me to be a more fundamental problem, however, referring to microeconomics as well. Since most readers will not agree, let me explain.

Economics is interested in general tendencies rather than in the idiosyncrasies of particular cases. Hence it aims to devise representative concepts from the beginning, i.e. modes of behaviour and circumstances elucidating general regularities rather than historical noise. This interest in general tendencies holds true for microeconomics, too, as long as we employ assumptions like profit maximization or utility maximization. We assume thereby that these motives are dominant and other possible motives and resulting modes of behaviour can be ignored. But this means, of course, that we do *not* start explicitly from all the particularities of individual cases. Rather we start from representative notions directly, and this involves the aggregation problem from the outset. Although we might feel that it is less important in microeconomics than in macroeconomics, it is still there. Consider the reasons for assuming profit maximization or utility maximization as dominant motives. We might be convinced that these motives explain competitively viable actions and are enforced upon the agents through competitive pres-

15 I have introduced the two concepts of "typical" and "representative" agents in order to stress this difference from the beginning. The distinction is also present in Marshall (p. 381, see also Bliss) but is somewhat blurred because he uses the term representative agent in both meanings, and this has been overlooked quite often. Clower (p. 290) writes, for instance, with regard to the concept of aggregate demand: "In short, Keynes either had a dual decision hypothesis at the back of his mind, or most of the General Theory is theoretical nonsense" and proceeds from that argument to introduce the Keynesian notions into microeconomics, initiating thereby the "Keynesian" microeconomic models which are intended to give microeconomic underpinnings to Keynesian macro theories. But a possible validity of the dual decision hypothesis on the macro level does not at all imply that it is also valid on the micro level. See also Hahn, p. 35.

sures[16]. This amounts to starting from properties of the competitive system in explaining individual actions rather than deriving these properties from individual actions stringently, and in this sense the argument can be interpreted as being derived from a macroeconomic, or even holistic, approach[17].

1.4 Isolation and Aggregation

In dealing with partial issues, economics presupposes isolation; in starting from representative or aggregate notions, it presupposes aggregation. These are not separate issues, however; rather aggregative analysis can be viewed as a special case in isolation in that it neglects certain influences, termed "structural", and uses the ceteris paribus clause to fix these influences in the same way as other data are fixed in partial analysis. This, at least, is the view proposed here.

The plan of the book is as follows. Chapter 2 discusses isolation, Chaps. 3 and 4 elaborate some consequences, and Chap. 5 discusses the aggregation problem by interpreting macroeconomics as employing an isolating approach.

16 See Alchian on that.

17 We shall come back to this and similar questions in Sect. 5.5.

2. On Isolation

> **The object of our analysis is, not to provide a machine, or method of blind manipulation, which will furnish an infallible answer, but to provide ourselves with an organised and orderly method of thinking out particular problems . . .**
>
> **John Maynard Keynes**

2.1 The Heuristic View

The isolating approach, or ceteris paribus analysis, is as old as economics. Alfred Marshall, who has developed it most succinctly, characterises it as follows.

> The forces to be dealt with (in economics) are however so numerous, that it is best to take a few at a time; and to work out a number of partial solutions as auxiliaries to our main study. Thus we begin by isolating the primary relations of supply, demand and price in regard to a particular commodity. We reduce to inaction all other forces by the phrase "other things being equal": We do not suppose that they are inert, but for the time we ignore their activity. This scientific device is a great deal older than science: it is the method by which, consciously or unconsciously, sensible men have dealt from time immemorial with every difficult problem of ordinary life (Marshall, p. xiii).

This view, even if not correct in that generality, seems to be vindicated within the economic compass: The isolating approach appears to be a characteristic feature of economic analysis – consciously or unconsciously. Even if some economists held other methodological views, their actual work has been ceteris paribus analysis. Keynes put it bluntly: "This is the nature of economic thinking" (Keynes (1), p. 297).

The isolating approach will be delineated in the following as a *heuristic procedure.* No attempt will be made to develop any methodology, i.e. that this and that are the criteria for any true scientific procedure and everything else is devoid of meaning.

It seems obvious to me that any method can have only heuristic value since it is bound to prove itself fruitful in dealing with its subject, and nothing can be prophesised here (Adorno (1), pp. 113, 118 ff.). To develop a method *without* any idea of its subject seems to imply drawing conclusions without any presuppositions, an attempt which is bound to fail.

Hence our ideas about the nature of the subject determine consciously or unconsciously our methods[1]. Since the nature of the subject is to be

1 Cf. also Hayek (2) p. 332.

clarified by our enquiry, our thoughts about the subject will yield retroactions changing our ideas and our methods. The choice of a certain method mirrors our expectations, our conjectures or our longings for certain regularities. We ought to be flexible here.

We shall exclude from the discussion two other important aspects of methodology pertaining not to the nature of its subject but to the conditions of social success of a theory: One the one hand a theory ought to correspond to our natural cognitive dispositions (the structure of our brain, the "hardware"); on the other hand it ought to comply to the social conditions influencing our thinking (the social programming of our thoughts and the encouragement or suppression of certain tendencies of thought within a social system). These are questions in physiology, psychology, and sociology. All curtailments of thought emerging here limit their own foundation. If a theory were developed, for instance, that all thinking is simply a correlate of extrascientific individual or group interests, this theory would have to be judged accordingly. If the theory were explicitly exempted from its own judgement, however, the traditional and unconditional claim of theory would be vindicated: The claim for truth. This claim – whether it can be vindicated or not – is the implicit premise of any discourse (Sebag, p. 9). Any discussion about it seems to be futile, therefore.

Sometimes the question of social success of a theory is posed in a normative spirit: What are the criteria of scientific procedure assuring intersubjective verification? This question will also be disregarded in the following. That it is not necessarily appropriate to the particular subject matter has been remarked already. Furthermore, the criteria developed in this vein seem to be highly arbitrary: If a theory fails to meet them it remains open whether this is due to shortcomings of the theory itself or whether the intellectual scope of the implicitly hypothesized average individual is too restricted (Adorno (2), pp. 49 ff.). For our purposes, therefore, it might be sufficient to adopt the position that a theory which has proved to be communicable has thereby cleared the hurdle of intersubjective verification. This implies nothing, of course, with respect to its intrinsic value or relevance which are always matters of experience; but this experience cannot be automatized by applying a set of formal rules.

Given these problems it seems wise to exclude methodological quarrels from our further discussion. The characterization of the isolating approach which we are going to propose is therefore without any methodological pretensions. We are simply leaning back, gaining some distance from economists' customary activities and a general view. The following discussion is meant to serve its purpose like a concert critique which we read in the morning paper and which discusses the previous night's performance in words rather than in music. Whilst raising points of agreement or disagreement, it leads us to realize our previous sensations and structural discernments more consciously and more deeply. Comparable to a concert critique, which makes sense only with reference to music, while losing all its meaning if taken in isolation, the following is an attempt to delineate some figures of economic thought in another language. [2]

Reverting to the old patterns of economic reasoning might entail some consequences apart from avoiding futile controversies based on mutual misunderstanding, however: A foundation for an understanding of the nature of macroeconomic laws – to be discussed in Chap. 5 – and a proper econometric modelling of ceteris paribus clauses – discussed in Chap. 4 – can among other things be provided thereby.

2.2 The Isolating Approach

2.2.1 The Marshallian Method

Marshall describes the isolating approach as follows.

> It is sometimes said that the laws of economics are 'hypothetical'. Of course, like every other science, it undertakes to study the effects which will be produced by certain causes, not absolutely, but subject to the condition that *other things are equal,* and that causes are able to work out their effects undisturbed. Almost every scientific doctrine, when carefully and formally stated, will be found to contain some proviso to the effect that other things are equal: The action of the causes in question is supposed to be isolated; certain effects are attributed to them, but only *on the hypothesis* that no cause is permitted to enter except those distinctly allowed for. It is true however that the condition that time must be allowed for causes to produce their effects is a source of great difficulty in economics. For meanwhile the material on which they work, and perhaps even the causes themselves, may have changed; and the tendencies which are being described will not have a sufficiently 'long run' in which to work themselves out fully. [3]

The ceteris paribus clause – the hypothetical premise that all other influences not explicitly mentioned are taken to be constant – isolates a certain set of relationships from its context. The explanation of the

---

2 Hence our considerations are concerned with the "rhetoric of economics" in the sense outlined by McCloskey.

3 Marshall p. 30. See also Marshall p. 304.

behaviour of certain variables starts from data which are generated by a ceteris paribus clause: everything other than the variables is assumed to be constant[4].

2.2.2 Substantive and Hypothetical Isolation

Disregarding the influence of disturbing causes by means of a ceteris paribus clause can be interpreted in two different ways:

In the first place, we might disregard those causes because they are unimportant. They do not significantly affect the relationships under study, and do not supersede or destroy them. If this holds true, we have performed what will be termed a *substantive isolation.*

In the second place we might use the ceteris paribus clause purely hypothetically, disregarding important influences for a while to narrow down the issue to a partial one which can be handled theoretically more easily. This might be a first step of an intellectual device to compose a coherent picture by later combining various partial answers so obtained. If we deduce the impact of a price change on demand by combining the income effect and the substitution effect, we combine two hypothetical effects to produce a substantive answer. While studying hypothetical effects, we have performed a *hypothetical isolation*[5]. We shall employ the same term irrespective of whether our hypothetical deductions are devised as building blocks for a theory of substantive contents, or are viewed as being of interest for their own sake[6].

---

4 Sometimes the constancy assumption refers not to the *level* of exogeneous variables themselves but to their *law of motion* which is taken to be constant (in growth theory, for instance). The *exogeneity* of data rather than their constancy is their essential feature. We can still proceed, however, to talk about constant data if we take the laws of motion, or the parameters describing them (growth rates for instance), as data rather then the exogeneous variables themselves. This might vindicate the simplification used henceforth in the text.

5 In Marshall, the distinction between these two types of isolation remains vague; he isolates in a substantivist spirit throughout, cf. Marshall, p. 304, for example. On the other hand, Schumpeter blurs the issue by maintaining: "This is always possible: anything can be labeled as a datum, which simply means that we give up the hunt for a purely economic explanation of whatever it is we so describe" (Schumpeter (1), p. 665). In another context he draws, however, a distinction between faulty and faultless isolations without saying what this means (Schumpeter (1) p. 538). The present discussion focuses on this distinction.

6 As a matter of terminology we shall employ the notions of abstraction and isolation synonymously in the following. (Abstracting from certain influenced means isolating a phenomenon in this regard from these influences.) Compare Schumpeter (1) p. 538 who uses the same terminology.

The distinction drawn here between substantive and hypothetical isolation is in practice not always a sharp one, of course. Rather it represents two polar cases with distinct and different implicit premises and purposes, and a different logic. The following sections will elaborate on these logical aspects. We do not preclude, however, these aspects being intermingled in actual work. Nevertheless it might be helpful to distinguish the two cases in order to adequately assess the meaning of particular segments of economic analysis.

2.3 Substantive Isolation

2.3.1 The Isolation Principle

In freezing some factors by means of the ceteris paribus clause, we transform them into data of our analysis. The choice of these data is not, however, simply a matter of caprice if the isolation is intended to be of substantive relevance. *A substantive isolation requires those factors transformed into data by means of the ceteris paribus clause have to be sufficiently stable with regard to the processes we want to explain such that the movements of the data do not destroy or supersede the relations we are studying in our model.* This is the *isolation principle.* If it is satisfied, the results obtained in a partial model will continue to hold approximately true even if we make allowance for these movements. If the isolation principle is not satisfied, however, we have abstracted from significant factors and will have obtained misleading results.

The isolation principle has two different aspects: temporal and causal isolation. The following sections deal with these two aspects.

2.3.2 Temporal Isolation

The problem of temporal isolation is emphasized by Marshall repeatedly. He stresses, for instance, that it "is true, however, that the condition that time must be allowed for causes to produce their effects is a source of great difficulty in economics"[7]. He refers here to the requirement that the adjustment of variables to the constellation determined by the data of the model be sufficiently fast in order that we may abstract from changes in the data.

7 Marshall p. 30. See also Marshall p. 304.

Consider for instance a partial model which gives the result that a vector x of certain variables converges to an equilibrium $\bar{x}$:

$$x \rightarrow \bar{x}$$

The equilibrium vector $\bar{x}$ is in turn determined by the data of the model. In particular, it is dependent upon those factors fixed under a ceteris paribus clause. These factors are actually changing, however, and this induces a movement of the equilibrium $\bar{x}$ over time. If this movement is sufficiently slow, the proposition that x tends towards $\bar{x}$ will remain approximately valid in the sense, for instance, that x tends towards $\bar{x}$ if it is not already very close to $\bar{x}$. If, on the other hand, the movement of the equilibrium values $\bar{x}$ is rather fast, the distance between x and $\bar{x}$ might even increase, the target $\bar{x}$ might move faster than the missile x. In that case the proposition that x tends to $\bar{x}$, which we have derived under the ceteris paribus clause, will actually be wrong and a tendency of x towards $\bar{x}$ cannot be isolated in a substantive sense.

More generally the *temporal isolation principle* can be formulated as follows: *A relationship is isolated in a temporal sense if the actual movement of the factors frozen under a ceteris paribus clause is not so fast that it destroys or supersedes the relationships derived in the model.* In short, the change of the data is required to be slow as compared to the speed of the changes in the variables under discussion.

2.3.3 Causal Isolation

A distinct, albeit not unrelated, aspect of isolation is the causal aspect. This is stressed by Eucken in a famous passage: "Economic data are those factors which determine the economic universe without being influenced directly by economic facts themselves."[8] With regard to our problem, and in dealing with partial analysis, we might rephrase this by saying that the factors taken as data of the analysis ought not be influenced directly by the processes or variables studied. If there are significant retroactions on the data of our model from the movements analyzed in that very model, the isolation becomes hypothetical. Since this holds true quite irrespective of whether the retroactions are direct or indirect, we shall speak henceforth of retroactions, and not solely of "direct retroactions".

---

8 Eucken p. 243, my translation.

Consider for instance a vector of variables x which has been shown to tend to an equilibrium vector $\bar{x}$ in a partial model:

$$x \to \bar{x}$$

The equilibrium $\bar{x}$ might depend in turn on the variables x directly or through their actual influence on other variables which we have fixed under a ceteris paribus clause, and the tendency of x to its equilibrium $\bar{x}$ might be destroyed. If, on the other hand, this retroaction is sufficiently weak, our proposition $x \to \bar{x}$ will be of substantive contents, and a *causal isolation* has been made.

More generally the causal isolation principle can be phrased as follows: *A set of relations is isolated in the causal sense if the actual retroactions of the movement of variables studied in the model on the data of the model, which have been generated by a ceteris paribus clause, is not so pronounced as to destroy or supersede the relationships derived in the partial model.* Put more bluntly, the retroactions of the variables on the data ought to be sufficiently weak.

2.4 Hypothetical Isolation

2.4.1 Analytical Isolation

In studying a complex problem it is, in Marshall's words, "necessary for man with his limited powers to go step by step; breaking up a complex question, studying one bit at a time, and at last combining his partial solutions into a more or less complete solution of the whole riddle."[9] In so doing, we might use the ceteris paribus clause, which generates the frame for each partial problem, solely for analytical convenience: We perform an analytical isolation without any intention of producing answers of substantive contents. We produce various theoretical components and aim to gain substantive implications when we combine them into an integral whole. Its sole purpose being simplification, the device of analytical isolation can be used arbitrarily.

9 Marshall p. 304.

2.4.2 Hypothetical Theories

We might, however, construct purely hypothetical theories for their own sake and without any attempt to use them later on as component parts of more realistic theories, as Marshall writes: "If we shut our eyes to realities we may construct an edifice of pure crystal by imaginations, that will throw side lights on real problems; and might conceivably be of interest to beings who had no economic problems at all like our own. Such playful excursions are often suggestive in unexpected ways: They afford good training to the mind: and seem to be productive only of good, so long as their purpose is clearly understood"[10].

We might, for instance, develop a theory of an imaginated barter economy while knowing for certain that a monetary economy works essentially differently. Nevertheless, the barter model, the edifice of crystal, might help us to understand much more deeply the crucially different features of a monetary economy. Similarly we might learn a lot about why our present institutions exist by analysing the difficulties of alternative hypothetical social arrangements such as labor management[11]. It might also be useful to develop purely hypothetical theories for other purposes: creating a theoretical or normative reference point, or handicrafting matchbox-sized models which permit us to analogize with regard to real economic problems. (If these analogies are thought to be a substantive rather than purely formal nature, we pass over from hypothetical to substantive isolation, of course.) Hence hypothetical theories might be quite useful – and even important – as long as their purpose is clearly understood.

2.4.3 A Remark on Scientific Discourse

In contrasting substantive and hypothetical isolation, we have tried to stress that the isolations and abstractions encountered in economics

10 Marshall p. 644.
Pigou sees things differently, however:
None the less, the thought-tools of the economist are, I think, in themselves and for their own sake of little interest and importance. The pure mathematician would protest – and rightly – if anyone should regard his structures as merely tools for physics and other applied sciences. But then these structures – if one who knows them by repute may venture to speak – constitute immense and imposing triumphs of the human intellect: they are themselves works of art. No claim of that kind can be made for the structure of pure economics. These are tools only. Those of them which cannot be made to work in elucidating the problems of the real world must be scrapped: there is no place for them in the gallery of art. But, though they are only tools, as tools they are vital. (Quotation taken from Lindahl p. 23 n.).

11 This is the characteristic feature of O. Williamson's "organizational failures approach".

differ in nature and purpose, this difference being perfectly legitimate so long as the particular purpose of each type of isolation is clearly understood.

It might be argued, however, that allowing for hypothetical theories implies that everything will be permitted to sail under the flag of economic science, the barest nonsense included, since every contribution can be vindicated as being either substantive or hypothetical. This plea is, however, theoretically misleading and practically the opposite of the truth.

The objection is theoretically misleading because a discrimination of theories according to their various possible purposes and claims requires that these claims can be named; and theories can be judged in a two-step procedure: First by evaluating to what extent these particular claims are met, and second by evaluating the relevance of the claims. The first step requires immanent criticism, provoking arguments as to why the claims are not met and what the particular shortcomings are, and counter arguments which might serve to delineate the meaning of the theory more sharply, extending or contracting its scope. Hence there is room for criticism and countercriticism.

Discussions of this type seem to be most fruitful since they tend to generate generalizations of existing and new theories. Apodictic programmes starting from a definition of some "true scientific procedure" will not. As Adorno says, "anything and nothing can be disproved from without"[12]. Or, in Marshall's words, there is a "general rule that in discussions on method and scope, a man is nearly sure to be right when affirming the usefulness of his own procedure, and wrong when denying that of others"[13].

Although we might still think the quality of a theory to be dependent on the relevance of its claims and on how far these claims are met, we will avoid futile controversies by bracketing out the problem that certain theories may have to be abandoned because they are unable to answer certain questions (which they were not intended to answer in the first place).

In practice, the banishment of hypothetical theories will not induce more relevant research but will vindicate possibly abstruse hypothetical theories. Criticizing a theory because it has no practical application and is hence hypothetical and irrelevant implies that all hypothetical theories are irrelevant, but such an assertion is felt by most theorists to

12 Adorno (1) p. 117.
13 Marshall p. 637 n 2.

be patently wrong. Hence this argument will not be convincing even if the theory under discussion is unreasonable in other respects[14]. Hence excluding hypothetical theories from "true science" will lead to a de facto immunization of these theories, and it seems more fruitful to focus the discussion on determining more exactly the scope, meaning and possible generalizations of the theories under consideration.

2.5 Economic Thinking

Economic thinking proceeds by imagination: By imagining economic processes, actions, and reactions we gain insight into their inner necessity. This is neither more subjective, nor more objective, than any other thinking.

Böhm-Bawerk has stressed this point in the context of capital theory superbly. He writes:

> I believe that the aberrations and confusions which we observe in this field are to a large extent due to an underestimation of certain difficulties which arise here, and to a resulting inattentiveness. To master the problems of capital theory involves – as anyone who has tried his skill here will feel – many deductive operations on imagined facts.
>
> And because so many factors are interacting here, the chains of deductive reasoning attain a precarious length. The dangers involved in these long chains of reasoning have been stressed most appropriately by Marshall, albeit partly for different reasons.
>
> I think our most dangerous enemy in those long chains of reasoning is the *word.* It ought to be a verbal veil only, suggesting vivid imaginations. But in how far we are following this suggestion, to what extent we are using our creative power to evoke the appropriate imagination, that is another question. But if we are somewhat careless, relying on the familiar word in guileless confidence, and are not creating the right imaginations in our mind, the word will serve as a veil in a different and undesired sense: It will veil the true contours of the economic ideas which we lose while being satisfied with the accidental silhouettes of the verbal image. And if these veiling words permeate long chains of reasoning, there is no assurance that the veiled imaginations with their unyielding features fit together. This might lead to dialectical conclusions drawn from strings of words, resulting in ominous dialectical derailments which we find in so many long-chained deductions but particularily in those heedless deductions encountered in capital theory[15].

---

14 Hence the many attacks on general equilibrium theory which elaborate on its supposedly purely hypothetical nature and conclude therefrom that it is useless altogether seem to cast away the good with the bad; the assaults by Galbraith and Kaldor might serve as examples here. These attacks remain therefore largely futile. The importance of general equilibrium in elucidating various issues seems to be patent: how else can one discuss the shortcomings of the price system with regard to externalities in a different framework, for instance?

15 Böhm-Bawerk p. Xff. In view of Böhm-Bawerk's beautiful German, I have been able to give here only a particularly inadequate translation. For the original, and a brief discussion of Böhm-Bawerk's notion of dialectics, see Schlicht (5), p. 47.

I am tempted to add that the mathematical symbol might be as dangerous as the word if it stands for purely mathematical rather than economic notions [16].

If we conceive economic theories without this background of economic imaginations, we will be unable to assess the significance of the underlying abstractions and isolations, and it will be practically impossible to appraise the significance and the empirical implications of those theories. No physicist would reject the law of the pendulum because the rope holding the pendulum happens to break. Rather he will know from his theoretical imagination that this accident constitutes a significant change in his test conditions, not an insignificant influence such as, for instance, a change in the number of persons watching the experiment. The problem of isolating essential aspects is similar in economics. To stress the isolation principle and the distinction between substantive and hypothetical isolation is to aim to conceive and delineate the meaning of economic theories with reference to broader (real or imagined) economic processes.

---

16 Keynes (1) p. 297 writes with respect to the hazards of mathematical symbolism:

It is a great fault of symbolic pseudo-mathematical methods of formalizing a system of economic analysis [...] that they expressly assume strict independence between the factors involved and lose all their cogency and authority if this hypothesis is disallowed; whereas, in ordinary discourse, where we are not blindly manipulating but know all the time what we are doing and what the words mean, we can keep 'at the back of our heads' the necessary reserves and qualifications and the adjustments which we shall have to make later on, in a way we cannot keep complicated partial differentials 'at the back' of several pages of algebra which assume that they all vanish.

This comment seems to me to correspond for mathematical symbols to what Böhm-Bawerk has said about words. Seen together, these two views imply that neither mathematical nor nonmathematical methods exclude 'blind manipulation' from the outset. In view of the transparency which can gained by using mathematical symbols, and which might be helpful in avoiding blind manipulation, Keynes' rash judgement seems to me to deserve qualification, however.

3. The Moving Equilibrium Method

> Statics is really but a branch of Dynamics, and partly because all suggestions as to economic rest, of which the hypothesis of a Stationary State is the chief, are merely provisional, used only to illustrate particular steps in the argument, and to be thrown aside when that is done.
>
> Alfred Marshall

3.1 Economic Equilibrium

Economic laws are, in Marshall's felicitous phrase, laws of tendencies: statements regarding the impact of various influences, more or less certain, more or less definite (Marshall p. 27). If these tendencies have worked themselves out fully, equilibrium is reached. Hence an equilibrium state is a state towards which things are tending. It might be, however, that an equilibrium is never reached since time is required for the causes to work out their effects: "For meanwhile the material on which they work, and perhaps even the causes themselves, may have changed; and the tendencies which are being described will not have a sufficiently 'long run' in which to work themselves out fully[1]." But even if this holds true and the moving equilibria are never reached, the notion of such equilibria might still be helpful to describe the tendencies at work. This is the theme of the present chapter.

The equilibrium notion underlying this discussion is the *Marshallian* one which has been sketched above. It is certainly only one of a host of various equilibrium notions used in economics. It is beyond the scope of the present considerations to offer a comprehensive discussion here, but a few remarks might be in place[2].

The *analytical* equilibrium notion refers to a rest point, which, if attained, will perpetuate. This notion is closely related to the Marshallian one: A Marshallian equilibrium is an analytical equilibrium which is stable[3].

The *Swedish* equilibrium notion refers to a state where expectations are fulfilled. A Swedish equilibrium can be an analytical (or Marshallian) disequilibrium since the state might change – even if correctly anticipated. This concept is important in microeconomics, particularly

1 Marshall p. 30.

2 Some references and a more detailed elaboration of the following remarks can be found in Schlicht (4).

3 See p. 30, n. 6 on the stability concepts used in this book.

in game theory where it is expected that everybody selects his best strategy given the best strategies of the others.

In macroeconomics the Swedish equilibrium notion poses the difficulty, however, of unclarity about what is meant by the expectations or plans of an aggregate, since the aggregate has no conscience of its own. Furthermore, we might like to talk about equilibria resulting from actions based on mutually inconsistent expectations – an equilibrium between bulls and bears, for instance – and here the Swedish notion is clearly inapplicable. The basic idea underlying this notion is, however, that a disappointment of expectations will lead to revisions of expectations and plans, and hence to changes; in that it boils down to the analytical notion. We might still want to speak, however, about a short-run equilibrium involving disappointed expectations, which induces changes in expectations leading to another short run equilibrium which might disappoint expectations again, and so forth. Here the Swedish notion poses problems, and it seems most appropriate therefore, to avoid it in our further discussion[4].

The *Walrasian* equilibrium notion refers to market clearance: A market is in Walrasian equilibrium if supply equals demand. A Walrasian equilibrium can be a Swedish disequilibrium, and vice versa, since agents might expect that a market will not clear; and it can be a Marshallian (or analytical) disequilibrium, and vice versa, if there are tendencies at work which produce uncleared markets. Furthermore it is restricted to models where supply and demand can be defined unambiguously – which is not the case if we allow for monopolies, buffer stocks and inventories, or other complications. In order to avoid these problems, we shall not employ the Walrasian notion either.

These brief remarks might encourage us in our reliance on the Marshallian and the analytical equilibrium notions: It is not maintained that other possible notions are useless, since they are certainly fruitful in many particular applications. Rather they are not sufficiently broad for our more general purposes. It is to be kept in mind, however, that the term "equilibrium" is fairly ambiguous as long as we do not make clear which particular notion we are employing, and that we shall stick to the Marshallian one.

---

4 The Hahn notion of equilibrium – namely "that an economy is in equilibrium when it generates messages which do not cause agents to change the theories they hold or the policies which they pursue" – poses the same difficulty. It improves upon the Swedish equilibrium notion, however, in that it makes clear what "disappointment of expectations" means if expectations assign probabilities to various possibilities rather than refer to one particular expected set of circumstances (see Hahn p. 25).

3.2. Moving Equilibria

3.2.1 Partial Analysis

Consider an economy and its laws of motion. Denote the *state* of the economy by s. For our purposes it is sufficient to take s as a real k vector. The laws of motion of the economy can be viewed as a set of differential equations associating to each state s the change of state $\dot{s}$ over time[5]

$$\dot{s} = \Psi(s) \quad s, \dot{s} \in \mathbb{R}^k. \tag{3.1}$$

If (3.1) represents the true economy, it will be very complex, and of high dimensionality, involving psychological, biological, meteorological, and physical laws. Hence we will be unable to write it down explicitly although we might still think that it can be conceived by a superior brain. Hence we are forced to do partial analysis.

In partial analysis, we proceed as follows. We split the vector s describing the state of the economy into a vector $x \in \mathbb{R}^m$ of *fast variables* and the vector $y \in \mathbb{R}^n$, $n = k - m$ of the remaining *slow variables*. Without loss of generality we can write the laws of motion given in (3.1) equivalently as

$$\dot{x} = f(x, y) \quad x, \dot{x} \in \mathbb{R}^m, \quad m < k. \tag{3.2}$$

$$\dot{y} = g(x, y) \quad y, \dot{y} \in \mathbb{R}^n, \quad n = k - m \tag{3.3}$$

and we can proceed to analyse the partial model (3.2) separately by fixing the slow variables y under a ceteris paribus clause. If we have labelled only a few variables as fast, the partial model might be rather simple and accessible to analytical treatment.

The conclusions drawn from the partial model will be of substantive relevance, we hope, if we can be sure that the speed of movement of the slow variables y (without, however knowing their law of motion) is in fact sufficiently slow as compared to the movements of the fast variables x. This is the requirement posed by the isolation principle in Chap. 2. Temporal isolation requires the movement of the slow variables to be slow, and causal isolation calls for a neglegible impact of the fast variables x on the movement of the slow variables y, i.e. on $g(x, y)$.

5 In order to avoid mathematical pirouettes, we assume all regularity and integrability conditions which might be required to be satisfied.

3.2.2 Comparative Statics

In comparative statics, we restrict our attention to the partial system (3.2) and denote, hence, the fast variables as *endogeneous* and the slow variables as *exogeneous*. Write the partial model (2) as

$$\dot{x} = f(x; y) \quad x, \dot{x} \in \mathbb{R}^m, y \in \mathbb{R}^n \tag{3.4}$$

where the semicolon indicates that the variables which follow are taken as fixed parameters.

Assume now that all solutions of (3.4) tend to a unique equilibrium

$$\bar{x} = \bar{x}(y) \quad \text{with} \quad f(\bar{x}(y); y) = 0 \tag{3.5}$$

i.e. $\bar{x}$ is stable[6].

If this equilibrium is approached rather quickly, we are entitled to approximate the true movement of the endogeneous variables by the movement of their equilibrium values $\bar{x}(y)$. We view the wanderings of the exogeneous (slow) variables as causes of the movement of the endogeneous (fast) variables, so to speak, and this can be achieved by studying the function $\bar{x}(y)$.

This is the method of comparative statics, which can be described in the words of Keynes as follows: "The object of a model is to segregate the semi-permanent or relatively constant factors from those which are

6 The concept of stability is used here and in the following in the sense of asymptotic stability for some region of attraction. The point (respectively set) $\bar{x}$ is asymptotically stable if for any solution $x(t)$ of (3.4) starting in the region of attraction the following holds true: (a) $x(t) \to \bar{x}$ for $t \to \infty$; (b) any neighborhood U of $\bar{x}$ contains a neighborhood V of $\bar{x}$ such that $x(t) \in V$ implies $x(t') \in V$ for all $t' \geqq t$. Refer to Bhatia/Szegö for more mathematical details.
To restrict the stability concept in such a way simplifies matters and offers the following advantage. Points which are stable but not asymptotically stable in the usual sense imply structural instability: Arbitrarily small changes in the system can change the qualitative behaviour of the solutions. This situation is clearly undesirable in economics, since economic models are never exact and the ceteris paribus clause will never be fully met. On the other hand, an asymptotically stable rest point implies (local) structural stability of the system (see Markus, theorems 1 and 4).
It is to be stressed that the term "stability" is used with two different meanings in this book: Referring to movements of *variables*, as in the present context, it pertains to movements of solutions of dynamical systems towards equilibrium; referring to *data* of the analysis, as in the foregoing chapter, it pertains to their constancy over time. I have decided to stick to this well-established, although ambiguous, usage of the term, rather than to introduce new notions, hoping that this will not lead to confusion. Note, however, that the term "structural stability", alluded to in Sect. 3.4.1 below, refers to still another stability concept (see Markus).

transitory or fluctuating so as to develop a logical way of thinking about the latter, and of understanding the time sequences to which they give rise in particular cases"[7]. Two additional remarks are in place here. Firstly it is not always necessary to start from an explicit law of motion (3.2). In many cases we might conjecture that possible alternative laws of motion in the partial system lead to the same equilibrium (3.5). If we have substantiated this conjecture, we are entitled to write down the equilibrium conditions (3.5) directly and start therefrom. As Malinvaud puts it:

To rely ... on a equilibrium formalization is to accept a short cut, i.e. the consideration of those equilibrium states that would result from dynamic adjustments.... The result of the analysis then depends only on the definition chosen for equilibrium and not on the precise specification of the dynamic process that is supposed to lead towards this equilibrium. One does not need to inquire precisely into this process in order to draw conclusions; in other words, these conclusions are supposed to be "robust" with respect to the formulation of the process (Malinvaud p. 7).

Secondly, the general idea developed here is not restricted to comparative statics in a narrow sense: Our partial model might generate cycles, for instance, and we might analyse their shape and position as determined by the values of the exogeneous variables. Hence a more comprehensive formalization of comparative statics, covering also the case of multiple equilibria, reads as follows: Assume there exists for any y a compact (i.e. closed and bounded) set $A(y) \subset \mathbb{R}^m$ such that all solutions of the partial system (3.4) tend to A and all solutions starting close to A remain close to A forever. This set $A(y)$ is named a *stable attractor* for system (3.4)[8]. If we want to stress that it is an attractor with respect to a partial system, we will use the term *partial attractor.* This attractor might contain, e.g., rest points (multiple equilibria), periodic points (limit cycles), but also various motions.

The comparative static method can be thought of as determining the correspondence $A(y)$, i.e. the partial attractor as determined by the state of the exogenous variables, and also as studying the impact of the exogenous variables on its shape and position. Examples for this kind of procedure can be found in business cycle theory, for instance. In Sect. 3.4 we give some mathematical underpinnings for this idea.

7 Keynes, (3) p. 296 f.

8 See footnote 6 p. 30.

3.2.3 The Moving Equilibrium Method[9]

Assume we have analysed the partial model (3.4) and have determined a unique equilibrium, i.e. the function $\bar{x}(y)$ in (3.5). The moving equilibrium method uses this result as a tool for simplifying the analysis of the complete model (3.2), (3.3) in the following manner: By assuming that a short run equilibrium always prevails, we can substitute x by $\bar{x}(y)$ in the equation $\dot{y} = g(x, y)$ describing the movements of the slow variables.

This substitution eliminates the fast variables and describes the movement of the slow variables as determined by their state alone: Instead of the true system of differential equations (3.2), (3.3) we obtain the differential equation

$$\dot{Y} = g(\bar{x}(Y), Y) \quad Y, \dot{Y} \in \mathbb{R}^n \tag{3.6}$$

which describes the evolution of the variable Y serving as an approximation for y.

This kind of approximation seems to be sensible as long as the isolation principle holds true for the partial system, i.e. as long as the method of comparative statics makes sense.

3.3 Extensions

3.3.1 The Marshallian Market[10]

The moving equilibrium method can be illustrated by means of the following example. Consider a market with a demand z as a linear and decreasing function of price p:

$$z = a - b \cdot p \quad a, b > 0 . \tag{3.7}$$

---

9 The moving equilibrium method has been used by Marshall most expediently. The mathematical treatment is due to Lotka (Chap. II). Samuelson, pp. 321–323, has introduced it into formal economics. It has been used widely in synergetics, recently, under the name "adiabatic elimination" (see Haken, Chap. 7). The present exposition follows Schlicht (2).

10 cf. Marshall p. 288 n; Schneider pp. 302–312; Allen (1) pp. 21 ff.

It is assumed that supply y is fixed in the short run. Hence, for a given price p, a certain excess demand $(z-y)$ will result, and this leads to changes in price: If excess demand is positive, the price will increase; if negative, the price will decrease:

$$\dot{p} = \lambda(z-y) \quad \lambda > 0. \tag{3.8}$$

By using (3.7), this can be written as

$$\dot{p} = \lambda(a - b \cdot p - y). \tag{3.9}$$

This is the equation describing the movement of price as a function of the prevailing price and current supply.

Supply y will change over time, however, according to the price which can be obtained

$$\dot{y} = \mu(p - c - d \cdot y) \quad \mu, c, d > 0. \tag{3.10}$$

This can be interpreted in the following way. The expression $(c + d \cdot y)$ denotes the Marshallian "supply price", i.e. the price which would lead to a permanent and constant supply of y. If the prevailing price exceeds the supply price, supply increases; if it is lower than the supply price, supply decreases. The schedule $p = c + dy$ is the long-run supply schedule. Short-run supply is perfectly inelastic.

The model (3.9), (3.10) can be treated by means of the moving equilibrium method if the speed of price adjustment λ is much faster than the speed of supply adjustment μ by taking price as the fast and supply as the slow variable.

The equilibrium price $\bar{p}(y)$ for given supply is determined by putting (3.9) to zero

$$\bar{p}(y) = (a-y)/b. \tag{3.11}$$

Inserting this into (3.10) yields the moving equilibrium approximation equation for the slow variable

$$\dot{Y} = \mu\{(a/b - c) - (d + 1/b)\, Y\}. \tag{3.12}$$

This differential equation has the general solution [11]

$$Y(t) = \bar{Y} + e^{\zeta t}(Y_0 - \bar{Y}). \tag{3.13}$$

11 An elementary introduction to differential equations can be found e.g. in Allen (1) Chap. 5.

Y_0 denotes the initial value of Y, $\bar{Y}$ denotes the equilibrium of (3.12) and ζ is the associated characteristic root:

$$\bar{Y} = (a - bc)/(1 + bd), \quad \zeta = -\mu(1 + 1/b). \tag{3.14}$$

Since $\zeta < 0$, the equilibrium is stable and the associated price is obtained by inserting $\bar{Y}$ into (3.11):

$$\bar{P} := p(\bar{Y}) = (c + da)/(1 + bd). \tag{3.15}$$

The solution of (3.13) can be compared now with the true solution of system (3.9), (3.10). This system has the characteristic roots

$$\zeta_{1/2} = \tfrac{1}{2}\{-(\lambda b + \mu d) \pm \sqrt{(\lambda b + \mu d)^2 - 4\lambda\mu(bd + 1)}\}. \tag{3.16}$$

Since the radicand is less than $(\lambda b + \mu d)^2$, both roots have negative real parts. Hence the stable equilibrium values are

$$\bar{p} = (c - da)/(1 + bd), \quad \bar{y} = (a - bc)/(1 + bd). \tag{3.17}$$

These are identical to those obtained by the moving equilibrium method.

The stability properties of the approximation turn out to be correct, too: it is stable. But consider more closely the dynamic behaviour of the moving equilibrium approximation as compared to the true movement.

Assume that prices are moving sufficiently fast as compared to supply changes to guarantee

$$\frac{\lambda}{\mu} > \frac{1}{b^2}\{bd + 2 + 2\sqrt{bd + 1}\} \tag{3.18}$$

which implies real roots

$$\zeta_2 < \zeta_1 < 0. \tag{3.19}$$

In particular we obtain for $\lambda \to \infty$

$$\lim_{\lambda \to \infty} \zeta_1 = \zeta, \quad \lim_{\lambda \to \infty} \zeta_2 = -\infty. \tag{3.20}$$

The general solution of (3.9), (3.10) can be written as

$$p(t) = \bar{p} + c_1 e^{\zeta_1 t} + c_2 e^{\zeta_2 t}$$

$$y(t) = \bar{y} + c_3 e^{\zeta_1 t} + c_4 e^{\zeta_2 t}$$

where c_1, c_2, c_3, c_4 are linear functions of $(p_0 - \bar{p})$ and $(y_0 - \bar{y})$ with p_0, y_0 as initial values.

If λ is large, $e^{\zeta_2 t}$ approaches zero very quickly and y approaches equilibrium with $e^{-\zeta_1 t}$. For large λ we have $\zeta \approx \zeta_1$ and the speed of adjustment obtained by the moving equilibrium method is approximately correct.

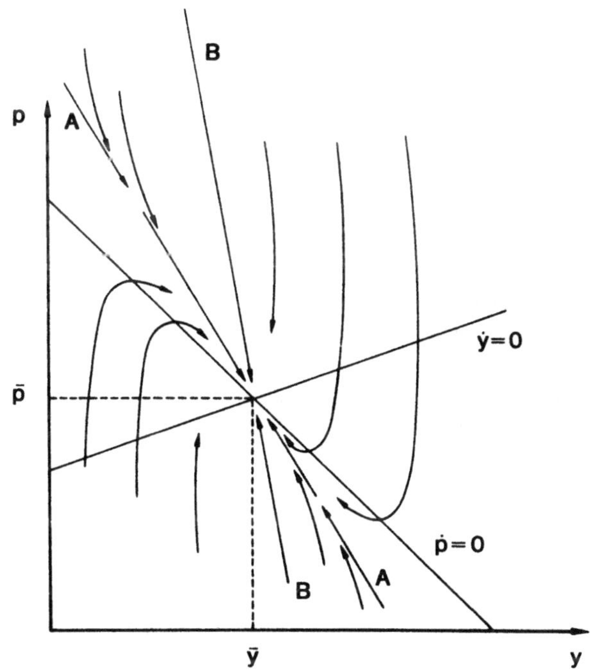

Fig. 1

In Fig. 1 the phase diagram of the true system (3.9, 3.10) is drawn. The curve marked by $\dot{p} = 0$ is obtained from putting $\dot{p}$ in (3.9) equal to zero, which gives $p = \bar{p}(y) = (a - y)/b$. Above this curve, p is decreasing; below the curve, it is increasing. Similarly the $\dot{y} = 0$ – locus is constructed by setting equation (3.10) equal to zero. To the left of this curve y rises, at the right it declines. In the upper orthant the trajectories move to the south-east; in the right orthant, to the south-west; in the lower orthant to the north-west; and in the left orthant to the north-east.

Furthermore we can look for those initial values (p_0, y_0) which lead to $c_2 = c_4 = 0$. These are on the locus A. It is the locus of those points where the speed of adjustment is determined solely by the first root ζ_1. Similarly the locus B is obtained by putting c_1 and c_3 equal to zero. On this locus it is the second root alone which describes the convergency towards equilibrium.

The equilibrium is a stable node. The trajectories are osculating with locus A first at the high speed ζ_2 and move close to this locus at the lower speed ζ_1. The larger λ, the more pronounced is the osculation with A. Furthermore it can be shown that A tends to the $\dot{p} = 0$ – locus for $\lambda \to \infty$, i.e. if price movements are very fast. The trajectories tend as straight lines towards A and continue along A to equilibrium [12].

3.3.2 A Qualification

The example might have illustrated the nature of the approximation obtained by the moving equilibrium method: It yielded the correct equilibrium and stability results. I would like to qualify the result, however, by pointing out that the speed of adjustment obtained by the moving equilibrium method is not necessarily the best approximation obtainable in the example.

As long as the speed of price adjustment λ remains finite, we have $\zeta \neq \zeta_1$. The speed of adjustment prevailing on the trajectory A, however, is exactly ζ_1, and along this locus the differential equation

$$\dot{y} = \zeta_1 (y - \bar{y}) \tag{3.21}$$

also holds true for the true system (3.9), (3.10): It gives the exact solution along A. On the other hand, the differential equation (3.9) implies

$$\dot{y} = \zeta (y - \bar{y}) \tag{3.22}$$

which holds true for no trajectory at all. Since all solutions are tending towards A and continue close to A towards equilibrium, it seems more

12 The proof can be sketched as follows. Define $z := (p - \bar{p})/(y - \bar{y})$ for $y \neq \bar{y}$. From (3.9) and (3.10) we obtain $\dot{z} = -(b \cdot z + 1) - \lambda(z^2 - dz)$. A and B are defined by the condition $\dot{z} = 0$ which leads to the corresponding two solutions z_1 and z_2. This implies $z_1 \to -b$ and $z_2 \to -\infty$ for $\lambda \to \infty$. Consider now a linear transformation of the vector (p, y) which transforms A into the abscissa and B into the ordinate. The transformed trajectories are given now by the family of functions $\xi = k\,|\eta|^{\lambda/\mu}$ generated by varying the parameter k. Hence the trajectories are approaching an angular shape for $\lambda \to \infty$ (cf. Erwe p. 138).

sensible to use the approximation (3.21) rather than the moving equilibrium approximation (3.22); intuitively it seems to be a better approximation in the sense that it describes the attracting trajectory A exactly.

For $\lambda \to \infty$, ζ tends towards ζ_1 and the approximations (3.21) and (3.22) coincide. As long as we are solely interested in qualitative results, however, we can still proceed to use the moving equilibrium method. That this holds true quite generally will be shown in the Sect. 3.4.

3.3.3 A Ramification

The moving equilibrium method can, in addition, be used iteratively: The vector Y can be partitioned again into "fast" and "slow" variables, and the moving equilibrium method can be used again to simplify the analysis of the moving equilibrium approximation (6), and so forth. This leads to a procedure, envisaged by Marshall, of starting with short-run analysis and proceeding, step by step, to the analysis of longer periods. This has been described by Samuelson as follows:

I, myself, find it convenient to visualize equilibrium processes of quite different speed, some very slow compared to others. Within each long run there is a shorter run, and within each shorter run there is a still shorter run, and so forth in an infinite regression. For analytic purposes it is often convenient to treat slow processes as data and concentrate upon the processes of interest. For example, in a short run study of the level of investment, income, and employment, it is often convenient to assume that the stock of capital is perfectly or sensibly fixed.

Of course, the stock of capital from a longer run point of view is simply the cumulation of net investment, and the reciprocal influence between capital and the other variables of the system is worthy of study for its own sake, both with respect to a hypothetical final equilibrium and the simple course of growth of the system over time.

So to speak, we are able by *ceteris paribus* assumptions to disregard the changes in variables subject to motions much "slower" than the ones under consideration; this is nothing but the "perturbation" technique of classical mechanics. At the same time we are able to abstract from the behavior of processes much "faster" than the ones under consideration, either by the assumption that they are rapidly damped and can be supposed to have worked out their effects, or by inclusion of them in the dynamical equations (derivatives, differences, etc.) which determine the behaviour of the system out of equilibrium.

The first of the above mentioned alternatives constitutes the justification for the use for comparative statics rather than explicit dynamics. If one can be sure that the system is stable and strongly damped, there is not great harm in neglecting to analyze the exact path from one equilibrium to another, and in taking refuge in a *mutatis mutandis* assumption. Of course, if one chooses to neglect certain dynamic processes, one may still retain others; e.g. in studying capital formation over two decades I may choose to neglect inventory fluctuations, but still may retain the acceleration principle in its secular aspects.

Under the second alternative where shorther run processes are contained in (say) the differential equations of the system, it is to be understood that these differential equations do not necessarily hold exactly at each instant of time. There may well be a still shorter run theory which explains how still higher differential equations lead to (rapidly) damped approaches to the postulated differential equation relations. And so forth in endless regression (Samuelson pp. 330 ff.).

This passage given an excellent characterization of the philosophy underlying the moving equilibrium method.

3.3.4 Discontinuities

Consider the partial system

$$\dot{x} = f(x; y) \quad x, y \in \mathbb{R} \tag{3.23}$$

and take the slow variables as parameters in our further discussion.

Now look at the partial attractor $A(y)$. A possible shape is depicted in Fig. 2. For y close to zero, $A(y)$ contains only one stable rest point. If y increases, the moving equilibrium will move along the heavily drawn branch although between y_1 and y_2 there is also a lower stable equilibrium on the lower heavily drawn branch of the graph. If y is increased beyond y_2, however, the upper equilibrium vanishes and the moving equilibrium jumps to the lower branch of the graph and continues to move along that branch if y is further increased.

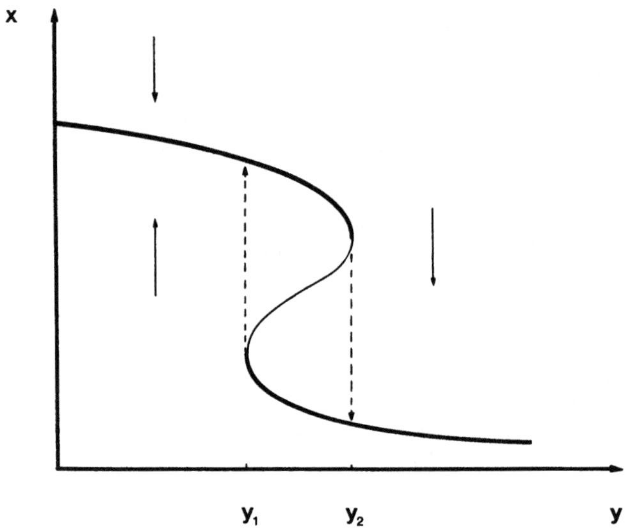

Fig. 2

If the process is reversed, and y is decreased, the corresponding moving equilibrium follows the lower branch till y_l and jumps to the upper branch if y is further decreased.

Hence an analysis of moving equilibria might quite naturally lead to an understanding of discontinuous movements. The idea was developed by Kaldor in 1940 in the context of business cycle theory, and has been subjected recently to a refined mathematical treatment in catastrophe theory (see Thom).

The argument is not restricted to the two-dimensional case, of course: These discontinuities will occur whenever the appropriate moving equilibrium moves discontinuously, and the moving equilibrium method is an aid in understanding those qualitative differences in speed associated with regular movements along the graph and "catastrophic" movements off the graph.

Perhaps one could think that nature works continuously since all developments proceed in finite time, but it is very fruitful to conceive the differences in speed between regular and catastrophic movements as qualitative differences, and it is not without irony that the moving equilibrium method, as developed by Marshall, offers a major theoretical tool for the study of discontinuities in spite of the fact that Marshall choose the motto: *"Natura non facit saltum"* for his book!

3.4 Mathematical Aspects

3.4.1 Approximation by Moving Equilibrium

In this section we shall give some mathematical underpinnings for the idea of using the moving equilibrium method for simplifying the analysis of systems of differential equations.

Consider the set of differential equations

$$\dot{x} = \alpha f(x, y) \quad \alpha > 0, x \in \mathbb{R}^m, y \in \mathbb{R}^n \tag{3.24}$$

$$\dot{y} = g(x, y) \tag{3.25}$$

$$\dot{Y} = G(Y) \quad Y \in \mathbb{R}^n \tag{3.26}$$

Equation (3.24) describes the movement of the fast variables, and Equation (3.25) describes the movement of the slow variables. The system (3.24), (3.25) is assumed to be a true system, and the system (3.26) describes the movement of the slow variables by the moving equilibrium approximation.

In Equation (3.24) we have inserted a parameter α describing the speed of the fast process, and we will be interested in the question whether (3.26) gives a good approximation for the movement of the slow variables if α is sufficiently high, i.e. if the fast variables are indeed sufficiently fast.

Theorem. Assume the following:

1. For any $y \in \mathbb{R}^n$ there exists a compact set $A(y) \subset \mathbb{R}^n$ and a continuously differentiable function $\varphi(x, y)$ such that

$$\varphi(x, y) > 0, \varphi_x(x, y) \cdot f(x, y) < 0 \quad \text{for} \quad x \notin A(y) \tag{3.27}$$

$$\varphi(x, y) = 0 \quad \text{for} \quad x \in A(y). \tag{3.28}$$

Hence $\varphi(x, y)$ is a partial Ljapunov function. Its existence implies that the set $A(y)$ is a globally asymptotically stable attractor for the partial system (3.24). All partial solutions tend to $A(y)$ and remain in $A(y)$ forever. (See Sect. 3.2.2 regarding the notion of a partial attractor.)

2. The functions g and G satisfy

$$G(y) = g(x, y) \text{ for all } x \in A(y) \text{ and all } y. \tag{3.29}$$

3. There exists a compact set $B \subset \mathbb{R}^m$ and a continuously differentiable function Φ such that

$$\Phi(Y) > 0, \Phi_y G(Y) < 0 \quad \text{for} \quad Y \notin B \tag{3.30}$$

$$\Phi(Y) = 0 \quad \text{for} \quad Y \in B. \tag{3.31}$$

Hence $\Phi(Y)$ is a Ljapunov function for system (3.26) establishing that B is a globally asymptotically stable attractor for this system.

There exists then for all sufficiently large α a globally stable compact attractor C_α of system (3.24), (3.25) which converges for $\alpha \to \infty$ to

$$C := \{(x, y) \mid y \in B, x \in A(y)\}. \tag{3.32}$$

In other words: The attractor obtained by the moving equilibrium method (3.26) approximates an attractor of the true system (3.24), (3.25) for sufficiently large α.

Proof.

1. First we prove

$$\varphi(x, y) = 0 \text{ implies } \varphi_x(x, y) = 0 \text{ and } \varphi_y(x, y) = 0. \tag{3.33}$$

Denote by x_1 the first component of (x, y) and write

$$h(x_1) := \varphi(x_1; x_2, \ldots, x_m, y_1, \ldots, y_m) \text{ for some given } x_2, \ldots, x_m, y. \tag{3.34}$$

Now consider an x_1 with $h(x_1) = 0$. Since φ is positive definite, h is positive definite. Hence $h(x_1 + \varepsilon) \geqq 0$ and $h(x_1 - \varepsilon) \geqq 0$ for all $\varepsilon > 0$. Since h is continuously differentiable and $h(x_1) = 0$ we have

$$\lim_{\varepsilon \to 0} \frac{h(x_1 + \varepsilon)}{\varepsilon} = \lim_{\varepsilon \to 0} \frac{h(x_1 - \varepsilon)}{\varepsilon} \tag{3.35}$$

and hence $h'(x_1) = -h'(x_1)$ which proves $h'(x_1) = 0$ for $h(x_1) = 0$.

This argument can be reiterated for all components of x and y, which proves the proposition.

The same argument leads to

$$\Phi(Y) = 0 \text{ implies } \Phi_y = 0. \tag{3.36}$$

2. Define the set

$$C_\alpha := \{(x, y) \mid \alpha \varphi_x(x, y) f(x, y) + \varphi_y(x, y)\, g(x, y) + \Phi_y(y)\, g(x, y) \geqq 0\}. \tag{3.37}$$

Since the functions involved in the definition are continuous, C_α is closed. We prove

$$C \subset C_\alpha. \tag{3.38}$$

To see this, take a $(x, y) \in C$. Since this implies $\varphi(x, y) = 0$ we have $\varphi_x = \varphi_y = 0$ by (33); since it implies $\Phi(y) = 0$, we have $\Phi_y = 0$ by (3.36). Hence the inequality in (3.37) is satisfied.

3. Next we prove $C_\alpha \to C$ for $\alpha \to \infty$. To see that, take any $(x, y) \notin C$. If $\varphi(x, y) > 0$, we have $(x, y) \notin C_\alpha$ for sufficiently large α. If $\varphi(x, y) = 0$, the two first terms in the definition (3.37) of C_α are equal to zero, and $G(y) = g(x, y)$ holds true. Here $(x, y) \notin C$ implies $y \notin B$ and hence $\Phi_y(y)\, G(y) < 0$ which violates the inequality in (3.37).
4. Consider the function

$$L_\alpha(x, y) := \begin{cases} \varphi(x, y) + \Phi(y) & \text{for } (x, y) \notin C_\alpha \\ 0 & \text{for } (x, y) \in C \end{cases} \tag{3.39}$$

which the time derivative

$$\dot{L}_\alpha = \alpha\, \varphi_x(x, y) f(x, y) + \varphi_y(x, y)\, g(x, y) + \Phi_y(y)\, g(x, y). \tag{3.40}$$

It is nonnegative and strictly positive outside C_α, and it is strictly decreasing outside of C_α. Hence it is a Ljapunov function for C_α. Furthermore C_α is compact for sufficiently large α. Hence the set C_α is globally asymptotically stable for sufficiently large α. This completes the proof.

3.4.2 Additional Observations

Corollary. If the true system (3.24), (3.25) is linear and both the partial system (3.24) and the moving equilibrium system (3.26) have unique stable equilibria $\bar{x}(y)$ and $\bar{Y}$, the system (3.24), (3.25) has the unique stable equilibrium $\bar{x}(\bar{Y})$, $\bar{Y}$ if α is sufficiently large[13].

Proof. We have

$$\dot{x} = \alpha(a + A x + B y),\ \bar{x}(y) = -A^{-1}(a + B y) \tag{3.41}$$

$$\dot{y} = c + C x + D y \tag{3.42}$$

$$\dot{Y} = c - CA^{-1} a + (D - CA^{-1} B)\, y,\ \bar{Y} = -(D - CA^{-1} B)^{-1}(c - CA^{-1} a). \tag{3.43}$$

Obviously we have $\dot{x} = 0$ for $x = \bar{x}(\bar{Y})$. Furthermore $x = \bar{x}(\bar{Y})$ and $y = \bar{Y}$ implies for (3.42)

$$\dot{y} = c - CA^{-1} a - (D - CA^{-1} B)(D - CA^{-1} B)^{-1}(c - CA^{-1} a) = 0. \tag{3.44}$$

Hence this is an equilibrium for the true system.

13 This theorem has been proved by another method in Schlicht (2).

It is well-known that there exists for any stable A a positive definite matrix M satisfying[14]

$$A' M + MA = -I. \tag{3.45}$$

(A is called stable iff all roots have negative real parts, i.e. iff the partial system is stable.)

Hence

$$\varphi(x; y) = (x - \bar{x}(y))' M (x - \bar{x}(y)) \tag{3.46}$$

is positive definite and

$$\varphi_x \cdot f_x = (x - \bar{x}(y))' (A' M + MA) (x - \bar{x}(y)) = -(x - \bar{x}(y))' (x - \bar{x}(y)) \tag{3.47}$$

is negative definite. Since φ is continuous, all requirements for φ in the Theorem are fulfilled.

Furthermore there exists a positive definite matrix N satisfying

$$(D - CA^{-1} B)' N + N (D - CA^{-1} B) = -I \tag{3.48}$$

provided the moving equilibrium system (3.43) is stable, and the associated Ljapunov function

$$\Phi(Y) = (Y - \bar{Y})' N (Y - \bar{Y}) \tag{3.49}$$

is positive definite with

$$\Phi_Y G_Y = -(Y - \bar{Y})' (Y - \bar{Y}) \tag{3.50}$$

negative definite.

Hence all requirements of the Theorem are met which proves stability of $\bar{x}(\bar{Y})$, $\bar{Y}$. Since this is a single point, it is also the smallest attractor of the true system. Hence the stable solution $x = \bar{x}(\bar{Y})$, $y = \bar{Y}$ of the true system is unique.

Remark 1. In the linear case, the Jacobian

$$J := \begin{pmatrix} \alpha A & \alpha B \\ C & D \end{pmatrix} \tag{3.51}$$

14 Cf. Bellman pp. 243–245.

describes the qualitative behaviour of the system (3.41), (3.42). Assume A to be nonsingular. Hence we can write

$$J := \begin{pmatrix} I & 0 \\ \frac{1}{\alpha} CA^{-1} & I \end{pmatrix} \begin{pmatrix} \alpha A & \alpha B \\ 0 & D - CA^{-1} B \end{pmatrix} \tag{3.52}$$

and this implies that the eigenvalues of J are approximately the eigenvalues of αA and $(D - CA^{-1} B)$, respectively, if α is sufficiently large. Since $(D - CA^{-1} B)$ is the Jacobian of the moving equilibrium approximation (3.43), the method will yield correct qualitative conclusions, in general, if the system (3.41), (3.42) is *structurally stable,* i.e. if small changes in the system leave the qualitative behaviour of the system unchanged[15].

Remark 2. In the example of Sect. 3.3.1 ("The Marshallian Market") the moving equilibrium method leads to the correct equilibrium as well as to the correct stability conclusion independently of the speeds of adjustments. This is not, however, always the case. Consider e.g. the differential equation system

$$\dot{x} = -x - y \tag{3.53}$$

$$\dot{y} = 3x + 2y \tag{3.54}$$

with the associated moving equilibrium equation

$$\dot{Y} = -Y. \tag{3.55}$$

Here the partial system as well as the moving equilibrium system are stable and heavily damped, but the true system (3.51), (3.52) is unstable, its trace being positive.

Remark 3. The Theorem requires continuously differentiable Ljapunov functions. This seems no severe restriction in view of the fact that stability implies in any case existence of continuous Ljapunov functions and is usually proved, in nonlinear applications, by means of Ljapunov functions which are typically piecewise continuous. This is, of course,

15 I.e. if all eigenvalues have non-zero real parts in the linear case (see Markus Theorem 4).

sufficient for the purposes of Theorem 1 as long as continuous differentiability on the boundary of the attractors is assured. On the other hand, in the linear – or linearizable – case, the existence of continuously differentiable Ljapunov functions poses no problem, as has been demonstrated in the Corollary.

Remark 4. Assume that the short-run system (3.24) produces a stable limit cycle with period $T(y)$. Hence the average of x over the cycle is

$$\tilde{x}(y) := \frac{1}{T(y)} \int_0^{T(y)} x(t; x, y)\, dt \tag{3.56}$$

where $x(t; x, y)$ denotes the solution of the partial system (3.24) for given y and the initial value x on the limit cycle. Similarly we might define

$$\tilde{G}(y) = \frac{1}{T(y)} \int_0^{T(y)} g(x(t; x, y), y)\, dt \tag{3.57}$$

$$\dot{Y} = \tilde{G}(Y) \tag{3.58}$$

as our moving equilibrium approximation. For $\alpha \to \infty$, the speed of convergence towards the limit cycle tends to be instantaneous and the period of the cycle tends to zero. Hence the movement of y will be approximated by (3.58) and the qualitative conclusions reached will be correct if the systems are structurally stable.

3.5 Hicks-d'Alembert's Principle

Consider a falling stone. As long as it has not reached the ground, it can be viewed as being in disequilibrium. Its motion is governed, however, by the law that the force of gravitation equals the inertia opposing its acceleration. In this sense, it is in moving equilibrium while falling. This is called d'Alembert's Principle in mechanics, and it holds true quite generally: Any motion can be viewed as balancing propelling and resisting forces – else the movement would be either faster or slower than it actually is.

The same holds true in economics: If there is no inertia, adjustment is instantaneous; if adjustment is not instantaneous, this can be attributed to adjustment costs, or searching costs, or other obstacles to instantaneous adjustment.

Hicks seems to have been the first to make this point with regard to economic analysis. In *Value and Capital* he writes:

So far as this limited sense of equilibrium is concerned it is true that we assume the economic system to be always in equilibrium. Nor is it unreasonable to do so. There is a sense in which current supplies and current demands are always equated in competitive conditions. Stocks may indeed be left in the shops unsold; but they are unsold because people prefer to take the chance of being able to sell them at a future date rather than cut prices in order to sell them now.... In this (analytically important) sense the economic system... can be taken to be always in equilibrium...[16]

Since Hicks first applied d'Alembert's Principle to economics, we will refer to it as "Hicks-d'Alembert's Principle".

This principle says that any state can be viewed as an equilibrium state by referring to suitable additional influences[17]. Hence the notions of equilibrium and disequilibrium are theoretical notions in the sense that they always pertain to a particular theory which neglects those additional influences and that it is inadmissible to discern equilibria or disequilibria empirically, i.e. without reference to a particular theory.

Furthermore the Hicks-d'Alembert's principle elucidates the relationship between statics and dynamics: All movements can be viewed as moving equilibria (although it is not necessary to do so), and in this sense, dynamics presupposes statics. On the other hand, static equilibria can be viewed as being generated by (infinitely fast) dynamical adjustments. In this sense, statics presupposes dynamics. Hence statics and dynamics are closely interlinked, and there is no gulf between them.

16 Hicks p. 131.

17 Note that the principle applies not only to the Marshallian but also to the Swedish equilibrium notion, even in the Hahn generalization (p. 28 n. above): If the "theory" of an agent denotes the way in which he handles information or if a "policy" denotes the way in which the agent reacts to new information, his theory and policy will not be affected by new information.

4. Econometric Implications

... the element of Time, which is the centre of the chief difficulty of almost every economic problem ...

Alfred Marshall

4.1 Linking Theory and Empirical Experience

If economics is compelled to deal only with partial models which are changing over time and subject to unspecified and inexact ceteris paribus clauses, then its relation with empirical experience is somewhat vague. This haziness should not lead us to the extremist view that theoretical economics develops mere models of thought which cannot be confronted with empirical experience and empirical research requires one to ignore all complications arising from wandering ceteris paribus clauses and moving models. We shall attempt, rather, to explain that the view of economics expounded above is a good guide for devising statistical methods appropriate to our theoretical questions, thus interlinking theory and empirical evidence. The isolation principle offers a view which connects theoretical argumentes and empirical investigations, so we are not forced to view theory as merely devising arguments which can only be used for empirical investigations if stripped from their proper theoretical context. The present chapter sets out to elucidate this link between theory and empirical evidence by means of examples, thus stressing the general point of view, rather than offering detailed solutions.

4.2 Invariances in Coefficients Versus Invariances in Their Stability

4.2.1 The Classical Invariance Premise

Many economic models start with the assumption that the coefficients describing the interaction of economic variables remain unchanged over time. This might be termed the *classical invariance premise.*

Let us consider an example. Consider a consumption function relating aggregate consumption C to aggregate income Y. Economic theory writes simply

$$C = F(Y) \quad O < F' < 1, F(0) > 0 \tag{4.1}$$

and supposes that such a function is given for the period under analysis: It is valid ceteris paribus and satisfies the isolation principle.

In order to determine the functions from real data, additional assumption have to be introduced. Usually one proceeds as follows[1]. It is assumed that the consumption function can be approximated linearly (or by some other analytic function – but let us stick to the simplest case):

$$C = \alpha + \beta Y \quad 0 < \beta < 1, \quad \alpha > 0. \tag{4.2}$$

In order to interpret (4.2) empirically, C and Y must be dated. Let C_t and Y_t denote consumption and income in period t, respectively, and write

$$C_t = \alpha + \beta_0 Y_t + \beta_1 Y_{t-1} + \ldots + \beta_s Y_{t-s}; \sum_{\tau=0}^{s} \beta_\tau = \beta, \ \beta_\tau \geqq 0 \quad \text{for all } \tau. \tag{4.3}$$

This equation takes into account that the impact of income on consumption does not work instantaneously but rather is distributed over s periods.

Since there are other influences on consumption which are fixed under a ceteris paribus clause in (4.1) but cannot be excluded in empirical applications, a random disturbance u_t is added which is intended to capture these additional influences. In the simplest case it is assumed that u_t is independently identically distributed with expectation zero and variance σ^2:

$$C_t = \alpha + \sum_{\tau=0}^{s} \beta_t Y_{t-s} + u_t; \quad E u_t = 0, E u_t^2 = \sigma^2, E u_t u_\tau = 0 \quad \text{for} \quad t \neq \tau. \tag{4.4}$$

This is a typical econometric translation of the theoretical relationship (4.1). Given the time series of consumption and income, the parameters $\alpha, \beta_0, \ldots \beta_s, \sigma^2$ can be estimated, and this gives an empirical consumption function.

---

1 See Schönfeld (1) pp. 2 ff.

The classical invariance premise is the assumption that the behavioural coefficients $\alpha, \beta_0, \ldots \beta_s$, which describe the interaction of the economic variables, are constant over time[2]. The impact of those factors covered by the ceteris paribus clause in (4.1) enters only through the disturbance term u_t and its parameter σ^2.

4.2.2 Economic Models as "Models of Thought"

Keynes has criticized this assumption of time invariant economic coefficients in a very extreme manner:

> It seems to me that economics is a branch of logic, a way of thinking ... (It) is a science of thinking in terms of models joined to the art of choosing models which are relevant to the contemporary world. It is compelled to be this because, unlike the typical natural science, the material to which it is applied is, in too many respects, not homogeneous through time. The object of a model is to segregate semi-permanent or relatively constant factors from those which are transitory or fluctuating so as to develop a logical way of thinking about the latter, and of understanding the time sequences to which they give rise in particular cases ....
>
> In chemistry and physics and in other natural sciences the object of experiment is to fill in the actual values of the various quantities and factors appearing in an equation or formula; and the work when done is once and for all. In economics this is not the case, and to convert a model into a quantitative formula is to destroy its usefulness as an instrument of thought ....
>
> To do so would make it useless as a model. For as soon as this is done, the model loses its generality and its value as a mode of thought.[3]

Hence, according to Keynes, a theoretical function like (4.1) is useful precisely because it is unspecified:

Since we cannot reasonably assume this function to be time invariant, there is no point in estimating it under that premise, but the function remains theoretically useful because we can deduce conclusions from it, such as the multiplier theorem, which are independent of its exact position. In theory we need not rely on the classical invariance premise, and presupposing it will lead to misleading results.

---

2 If time dependent coefficients are explained by a function like $\alpha_t = a + bt$, the classical invariance premise pertains to the parameters a and b. The classical invariance premise is also present in Bayesian models since the coefficients, although seen as subjective random variables, are conceived of as being time invariant (see Schönfeld (2) Chap. 13).

3 Keynes (3) pp. 296 ff., 299, 296. See also Keynes (3) pp. 285–332 and Keynes (1) pp. 44, 297 ff.

4.2.3. Short-Run Invariances

Tinbergen's reply to Keynes' objection is simple and convincing: "Coefficients changing just by chance would, of course, render the whole of quantitative economic science impossible"[4]. We cannot identify a law from real data which changes erratically over space and time, and it is preferable to avoid the term under these circumstances. But economics is, in fact, interested in laws exhibiting regularities and connexions among similar but distinct events. Hence economics actually looks for invariances, but where?

The isolation principle provides a hint here: Economic models are conceived in such a way that the data of the model are stable with regard to movements of the variables under study. Hence we are entitled to assume that the economic relations we want to describe are *stable in the short run.* On this, Keynes and Tinbergen agree. Keynes writes: "... it is only in a short series ... that there is a reasonable expectation that the coefficients will be fairly constant."[5] And Tinbergen says: "In most cases only small changes in structure will occur in the near future."[6] In fact, as long as we presuppose a model to be substantively isolated, it cannot be a mere "model of thought" with no empirical implications, since hypotheses about relative stabilities enter by necessity.

But as to the implications, Tinbergen and Keynes part company. Whereas Tinbergen argues that this stability in a short series renders it plausible to employ the classical invariance premise and use the assumption of constant coefficients as an approximation, Keynes correctly points out: "One of the chief dilemmas facing you is, of course, ... that the method requires not too short a series whereas it is ony in a short series, in most cases, that there is a reasonable expectation that the coefficients will be fairly constant"[7].

Hence, even if we take short-run invariances as an acceptable assumption, the issue remains: Is there any practicable alternative avoiding both Keynes' "modell of thought" nihilism and the short cut provided by the classical invariance premise?

Note that the research programme of econometrics, which starts from the classical invariance premise, has not been particularly success-

4 Keynes (3) p. 292.
5 Keynes (3) p. 294.
6 Tinbergen p. 152.
7 Keynes (3) p. 294.

ful up to now. Simple extrapolation methods perform quite often better in the sense of producing less misleading predictions[8].

As Hayek remarked, the problems with econometrics cannot be attributed simply to the fact that econometrics is a rather young discipline: "Sir William Petty, the founder of Econometrics, was after all a somewhat senior colleague of Sir Isaac Newton in the Royal Society!"[9]

As the criticism of econometrics put forward by economists like Keynes, Hayek, or Schumpeter is directed mainly against the classical invariance premise, it might be fruitful to look for an alternative[10].

4.2.4 Invariances in Stabilities

So let us look for an alternative to the classical invariance premise. We do not want to assume time-invariant coefficients. Rather we would like to have slowly but unpredictably varying coefficients. Hence we take the vector of coefficients of our model at time t as a random vector and denote it by a_t. If a_t is assumed to change slowly over time, the conditional density $f(a_t|a_{t-1})$ will concentrate close to a_{t-1}. This amounts to the assumption of short-run stability, which is all we know from the isolation principle, and is certainly not enough for empirical analysis. We need additional strong assumptions.

The assumption involved in the classical invariance principle, namely that $a_t = a_{t-1}$ with probability one, is too strong, however, since it transforms short-run stability into long-run stability. This unwarranted implication can be avoided by relaxing it in assuming that the expectation of a_t equals a_{t-1}, and that a_t is distributed around a_{t-1} according to a certain time-invariant distribution. In other words, a_t is assumed to be generated by a random walk

$$a_t = a_{t-1} + v_t \tag{4.5}$$

involving the random variable v_t with $Ev_t = 0$ and a certain time-invariant distribution. The size of the covariance matrix $\Sigma := Ev_t v_t'$ measures the stability of the coefficients over time, and the classical

8 See Bhattacharyya and the references cited therein.

9 Hayek (3) p. 437.

10 It is interesting to note that Hayek and Schumpeter, while criticizing Keynes' macroeconomics, are actually criticizing the classical invariance premise: Their main point is always that time-invariant macroeconomic relations cannot be postulated. See on this point Hayek (2) pp. 91 ff., (3) p. 434; Schumpeter (1) pp. 91 ff., 473, 1170–1184, (2).

invariance premise is covered as the limiting case $\Sigma \to 0$. All this amounts to replacing the assumption of invariant coefficients by the assumption that the *stability* of the coefficients remains invariant over time. This *stability invariance premise* allows for both stability in the short run, as required by the isolation principle, and long-run instability of the coefficients at the same time, and is, hence, less vulnerable to the criticism raised against the classical invariance premise.

One could argue, of course, along similar lines against the assumption of time-invariant stabilities, but note that some assumption of this sort is implicitly made by maintaining that any theoretical model remains true over time: This implies that the data of the model remain sufficiently stable. In view of the fact that we have to assume invariances somewhere, the stability invariance premise seems to me to suggest itself as the most appropriate hypothesis[11].

4.3 Estimation in a Linear Model with Random Walk Coefficients

4.3.1 The Model[12]

In the following a simple linear econometric model is analysed which differs from the classical linear regression model only in the assumption that the coefficients are generated by a simple random walk. The discussion is intended to demonstrate that the stability invariance premise might lead, at least in a standard case, to statistically treatable econometric models. It is, however, beyond my scope to go into various generalizations and complications.

---

11 From a mathematical point of view it is not particularly difficult to replace (4.5) by the generalization

$$a_t = A a_{t-1} + v_t$$

but this requires additional information for determining the matrix A and runs into the problem that a_t might lead to probabilistic stationary states for $t \to \infty$, whereas the process (4.5) is really dissipative and uninformative in the limit. This seems to me to be more appropriate for economic models, which are not devised for giving a definite answer to problems in the indefinite future.

12 See also Athans, Cooley/Prescott, Pagan, and Hatanaka/Tanaka for analyses of related issues.

Consider the regression model

$$y_t = a_t' x_t + u_t, \quad t = 1, 2, \ldots, T, u_t \sim \mathcal{N}(0, \sigma^2) \tag{4.6}$$

with $y_t \in \mathbb{R}$, $x_t \in \mathbb{R}^n$ observations, $a_t \in \mathbb{R}^n$ coefficients to be estimated, and $u_t \in \mathbb{R}$ normal disturbances for the time periods $t = 1, 2, \ldots, T$. The coefficients are assumed to be generated by a random walk with normal disturbances $v_t \in \mathbb{R}^n$:

$$a_t = a_{t-1} + v_t, \quad t = 2, 3, \ldots, T; \quad v_t \sim \mathcal{N}(0, \Sigma). \tag{4.7}$$

The estimation problem can be discussed in two steps. First, the time path $a_1, a_2, \ldots, a_T$ of the coefficients is estimated under the assumption that the variances σ^2 and Σ are known. The proposed procedure will be developed in Sects. 4.3.2–4.3.4 and amounts to minimizing the weighted sum of the squares of the disturbances u_t and v_t, i.e.[13]

$$Q = \sigma^{-2} \sum_{t=1}^{T} u_t^2 + \sum_{t=2}^{T} v_t' \Sigma^{-1} v_t .$$

In Sect. 4.3.5, the estimation problem for the variances is discussed under the simplifying assumption that the changes in the coefficients are uncorrelated; Sect. 4.3.6 gives some approximation results under the assumption that the coefficients are sufficiently stable over time, as required by the isolation principle. Under these assumptions, the estimation procedure boils down to elementary matrix operations without inversions or other complicated calculations.

---

13 A straightforward way to estimate the time path $a_1, a_2, \ldots, a_T$ of the coefficients is to apply Kalman-Bucy filtering, and this has been the usual strategy. This way of dealing with the problem has severe shortcomings, however: First and foremost, Kalman filters are one-sided filters and are hence not optimal for the estimation of past coefficients, for which a two-sided filter ought to be used. Second, there exists no constant steady-state gain for updating the estimates, since the x_t's are time dependent. Hence the main computational advantage of Kalman-Bucy filtering is lost. Third, the estimation of initial values poses problems in the Kalman framework. The aim of Sect. 4.3 is to hint at an alternative procedure for estimating the coefficients optimally in a one-shot procedure without having recourse to initial values.

4.3.2 Notation

Define

$$y := \begin{pmatrix} y_1 \\ y_2 \\ \vdots \\ y_T \end{pmatrix}, \quad u := \begin{pmatrix} u_1 \\ u_2 \\ \vdots \\ u_T \end{pmatrix}, \quad a := \begin{pmatrix} a_1 \\ a_2 \\ \vdots \\ a_T \end{pmatrix}, \quad v := \begin{pmatrix} v_2 \\ v_3 \\ \vdots \\ v_T \end{pmatrix} \tag{4.8}$$

order $\quad T \times 1 \quad\quad T \times 1 \quad\quad Tn \times 1 \quad\quad (T-1)\, n \times 1$

$$X := \begin{pmatrix} x_1' & & & 0 \\ & x_2' & & \\ & & \ddots & \\ 0 & & & x_T' \end{pmatrix}, \quad P := \begin{pmatrix} -I & I & & & 0 \\ & -I & I & & \\ & & \ddots & \ddots & \\ 0 & & & -I & I \end{pmatrix} \tag{4.9}$$

order $\quad T \times Tn \quad\quad (T-1)\, n \times Tn$

and write (4.6), (4.7) as

$$y = Xa + u, \quad u \sim \mathcal{N}(0, \sigma^2 I) \tag{4.10}$$

$$Pa = v, \quad v \sim \mathcal{N}(0, S_v), \quad S_v := I \otimes \Sigma. \tag{4.11}$$

Observations y, X and variances σ^2, Σ are assumed to be known and the problem is to estimate therefrom the timepath of the coefficients a.

A descriptive approach would be to minimize the sum of squares, i.e.

$$Q = \sigma^{-2} u' u + v' S_v^{-1} v. \tag{4.12}$$

A necessary condition for a minimum is obtained by inserting (4.10) and (4.11) into (4.12) and putting the derivatives to zero

$$(X' X + \sigma^2 P' S_v^{-1} P)\, a = X' y. \tag{4.13}$$

A singularity problem will arise only for matrices X satisfying $x_1' c = x_2' c = \ldots = x_T' c$ for some nonzero $n \times 1$ vector c and can, hence, be ignored: The least-squares solution can be taken to be unique.

4.3.3 A Parametrization

Consider the set of the n orthogonal solutions z to $Pz = 0$. These give rise to an orthogonal matrix Z of order $Tn \times n$ with

$$PZ = 0, \quad Z'Z = I, \quad P'(PP')^{-1}P + ZZ' = I. \tag{4.14}$$

Any solution to (4.11) can be written as

$$a = P'(PP')^{-1}v + Z\lambda, \quad \lambda \in \mathbb{R}^n \tag{4.15}$$

where λ stands for a parameter vector.

Inserting (4.15) into (4.10) gives rise to the regression

$$y = XZ\lambda + w, \quad w := XP'(PP')^{-1}v + u \tag{4.16}$$

for the unknown parameters λ. The disturbances w are normally distributed:

$$w \sim \mathcal{N}(0, S_w), \qquad S_w := XP'(PP')^{-1}S_v(PP')^{-1}PX' + \sigma^2 I. \tag{4.17}$$

The maximum likelihood (Aitken) estimate $\hat{\lambda}$ satisfies

$$Z'X'S_w^{-1}(y - XZ\hat{\lambda}) = 0. \tag{4.18}$$

4.3.4 Estimating the Coefficients

For given λ, y, X the system (4.15), (4.16) defines the distribution of a:

$$\begin{aligned}
&a \sim \mathcal{N}(Z\lambda + BS_w^{-1}(y - XZ\lambda), S_a),\\
&B := P'(PP')^{-1}S_v(PP')^{-1}PX'\\
&S_a := P'(PP')^{-1}S_v(PP')^{-1} - BS_w^{-1}B'.
\end{aligned} \tag{4.19}$$

Hence a maximum likelihood estimate for a can be derived by taking the mode of (4.19) for any $\hat{\lambda}$ satisfying (4.18):

$$\hat{a} := Z\hat{\lambda} + BS_w^{-1}(y - XZ\hat{\lambda}). \tag{4.20}$$

Theorem. If $\hat{\lambda}$ satisfies (4.18), $\hat{a}$ from (4.20) will satisfy (4.13). Hence the minimization of the sum of squares (4.12) gives a maximum likelihood estimate for the coefficients a.

Proof. Premultiplication of (4.20) by $(X'X+\sigma^2 P' S_v^{-1} P)$ will result in $X'y$:

$$
\begin{aligned}
(XX'+\sigma^2 P' S_v^{-1} P)\,\hat{a} = {} & (X'X+\sigma^2 P' S_v^{-1} P)\, Z\hat{\lambda} \\
& + X'XBS_w^{-1}(y-XZ\hat{\lambda}) \\
& + \sigma^2 P' S_v^{-1} PBS_w^{-1}(y-XZ\hat{\lambda}) .
\end{aligned}
\tag{4.21}
$$

Since $PZ=0$, the first term reduces to $X'XZ\hat{\lambda}$. Since $XBS_w^{-1}=I-\sigma^2 S_w^{-1}$, the second term reduces to

$$X'y-\sigma^2 X' S_w^{-1} y - X'XZ\hat{\lambda}+\sigma^2 X' S_w^{-1} XZ\hat{\lambda} .$$

Using (4.14), the third term reduces to

$$\sigma^2 (I-ZZ')\, X' S_w^{-1}(y-XZ\hat{\lambda}) .$$

In view of (4.18) this reduces further to

$$\sigma^2 X' S_w^{-1} y - \sigma^2 X' S_w^{-1} XZ\hat{\lambda} .$$

Hence $X'y$ remains and all other terms cancel, Q.E.D.

The accuracy of the estimate is described by the variance-covariance matrix S_a. Note that S_a, as defined in (4.19), is independent of λ.

4.3.5 Estimation of Variances

In this section we shall make the following simplifying assumption

Assumption. We assume that

$$\Sigma = \begin{pmatrix} \sigma_1^2 & & 0 \\ & \ddots & \\ 0 & & \sigma_n^2 \end{pmatrix} \tag{4.22}$$

is diagonal.

From (4.16) and (4.17) we get the likelihood function

$$L = -\log\det S_w - w' S_w^{-1} w . \tag{4.23}$$

For $\lambda = \hat{\lambda}$ and $a = \hat{a}$, the concentrated likelihood function can be derived: Define

$$\hat{u} = y - X\hat{a}, \quad \hat{v} = P\hat{a}, \quad \hat{w} = XP'(PP')^{-1}\hat{v} + \hat{u}. \tag{4.24}$$

Together with (4.16), (4.18), and (4.20) we get

$$\hat{u} = (I - XBS_w^{-1})(y - XZ\hat{\lambda}) = \sigma^2 S_w^{-1}\hat{w} \tag{4.25}$$

and hence

$$\hat{w}' S_w^{-1}\hat{w} = \frac{1}{\sigma^4}\hat{u}' S_w \hat{u}. \tag{4.26}$$

Using the definition of S_w and the relation $X'\hat{u} = \sigma^2 P' S_v^{-1}\hat{v}$ from (4.13), (4.24), this reduces to

$$\hat{w}' S_w^{-1}\hat{w} = \frac{1}{\sigma^2}\hat{u}'\hat{u} + \hat{v}' S_v^{-1}\hat{v}. \tag{4.27}$$

Hence the second term in the concentrated likelihood function associated with (4.23) turns out to be the minimum of the sum of squares Q as defined in (4.12). Using (4.13), this can be written as

$$\hat{Q} = \frac{1}{\sigma^2}(y'y - y'X(X'X + \sigma^2 P' S_v^{-1} P)^{-1} X'y) \tag{4.28}$$

and the likelihood function to be maximized is

$$L(\sigma^2, \sigma_1^2, \ldots \sigma_n^2) = -\log\det S_w - \hat{Q}. \tag{4.29}$$

From (4.28) we find

$$\frac{\partial\hat{Q}}{\partial\sigma^2} = -\frac{1}{\sigma^4}\hat{u}'\hat{u}; \quad \frac{\partial\hat{Q}}{\partial\sigma_i^2} = -\frac{1}{\sigma_i^4}\sum_{t=2}^{T}\hat{v}_{it}^2 \tag{4.30}$$

where $\hat{v}_{it}$ denotes the i-th component of $\hat{v}_t$.

The matrix S_w can be evaluated by defining the coefficients

$$r_{t,\tau} := \frac{1}{3}T + \frac{1}{2} + \frac{1}{6}\frac{1}{T} + \frac{1}{2}\cdot\frac{1}{T}(t^2 - t + \tau^2 - \tau) - \max(t, \tau)$$

$$t, \tau = 1, 2, \ldots, T \tag{4.31}$$

and the $T \times T$ matrix

$$H := (r_{t,\tau} \cdot x_t' \Sigma x_\tau), \qquad t, \tau = 1, 2, \ldots, T. \tag{4.32}$$

This permits us to write [14]

$$S_w \equiv \sigma^2 I + H. \tag{4.33}$$

Write furthermore

$$H_i = \frac{\partial H}{\partial \sigma_i^2} = (r_{t,\tau} \cdot x_{it} x_{i\tau}) \tag{4.35}$$

where x_{it} denotes the i-th component of x_t. With these expressions we have

$$\frac{\partial \log \det S_w}{\partial \sigma^2} = \operatorname{tr}(\sigma^2 I + H)^{-1} \tag{4.36}$$

$$\frac{\partial \log \det S_w}{\partial \sigma_i^2} = \operatorname{tr}(\sigma^2 I + H)^{-1} H_i \tag{4.37}$$

and we can determine the gradient of the concentrated likelihood function (4.29)

$$\frac{\partial L}{\partial \sigma^2} = -\operatorname{tr}(\sigma^2 I + H)^{-1} + \frac{1}{\sigma^4} \hat{u}' \hat{u} \tag{4.38}$$

$$\frac{\partial L}{\partial \sigma_i^2} = -\operatorname{tr}(\sigma^2 I + H)^{-1} H_i + \frac{1}{\sigma_i^4} \sum_{t=2}^{T} \hat{v}_{it}^2 \tag{4.39}$$

and the maximum can be searched by a gradient process.

4.3.6 An Approximation

The dimensionality of the problem seems to be amazing. The matrix

$$M := (X' X + \sigma^2 P' S_v^{-1} P) \tag{4.40}$$

---

14 For A as defined in (4.45) above, we have $P = A \otimes I$. The $r_{t,\tau}$ are the elements of

$$R := A'(AA')^{-1}(AA')^{-1} A.$$

appearing in the normal equation (4.13) is, however, definite and of rather simple band structure:

$$M=\left(\begin{pmatrix} x_1x_1' & & 0 \\ & x_2x_2' & \\ & & \ddots & \\ 0 & & & x_Tx_T' \end{pmatrix}+\sigma^2\begin{pmatrix} \Sigma^{-1} & -\Sigma^{-1} & & & 0 \\ -\Sigma^{-1} & 2\Sigma^{-1} & -\Sigma^{-1} & & \\ & -\Sigma^{-1} & 2\Sigma^{-1} & \ddots & \\ & & \ddots & \ddots & \\ & & & 2\Sigma^{-1} & -\Sigma^{-1} \\ 0 & & & -\Sigma^{-1} & \Sigma^{-1} \end{pmatrix}\right) \tag{4.41}$$

Hence a numerical solution does not seem to be out of reach, especially if Σ is diagonal.

The estimation of the variances seems to be rather complicated, too. If the coefficients are sufficiently stable as suggested by the isolation principle, we can assume $\sigma^2 \gg \sigma_i^2$, for all i, and can use the approximation

$$(\sigma^2 I+H)^{-1} \approx \frac{1}{\sigma^2} I - \frac{1}{\sigma^4} H \tag{4.42}$$

which will simplify the computation of (4.38) and (4.39) considerably.

Under the assumptions $\sigma^2 \gg \sigma_i^2$, T large, we can approximate the normal equation in a similar fashion. By maximizing $Q+v_2'v_2+v_T'v_T$ rather than Q we obtain the normal equation

$$(I+CX'X)\,a=CX'y \tag{4.43}$$

with

$$C:=\frac{1}{\sigma^2}((AA')^{-1}\otimes\Sigma) \tag{4.44}$$

where

$$A:=\begin{pmatrix} -1, & 1 & & & 0 \\ & -1, & 1 & & \\ & & \ddots & \ddots & \\ 0 & & & -1, & 1 \end{pmatrix} \quad \text{of order } T\times(T+1)\,. \tag{4.45}$$

Hence the elements of $(AA')^{-1}$ are simply

$$(AA')^{-1}=\left(\frac{1}{T+1}\min(t,\tau)\,(T+1-\max(t,\tau))\right) \quad t,\tau=1,2,\ldots. \tag{4.46}$$

For $\sigma^2 \gg \sigma_i^2$, for all i, we can approximate

$$(I+CX'X)^{-1}\approx I-CX'X \tag{4.47}$$

which implies

$$a \approx CX'y - CX'XCX'y. \tag{4.48}$$

For $\sigma^2 \gg \sigma_i^2$ we can also use the approximation (4.42) also in order to derive an approximation for the covariance matrix S_a from (4.19), which describes the accuracy of our estimates. In view of (4.11), (4.17), and (4.34) we can write

$$S_a \approx (R \otimes \Sigma) - (R \otimes \Sigma) X' \left(\frac{1}{\sigma^2} I - \frac{1}{\sigma^4} H\right) X (R \otimes \Sigma) \tag{4.49}$$

with R as given by (4.31), (4.33).

Note that (4.48) and (4.49) can be computed by elementary matrix operations, and inversions are avoided. The same holds true for the gradients of the likelihood function (4.38), (4.39). Hence the dimensionality problem does not infringe on computability substantially.

4.4 Unsettled Problems

In the preceding chapters it has been argued that economic laws, since conceived ceteris paribus, should be viewed as changing over time. The estimation procedure which has been explained in the preceding section is intended to serve as an example illustrating to what kind of considerations the proposed view of economic laws might lead, and that workable estimation techniques can be devised along the lines indicated by the isolation principle.

It is, however, an example dealing only with the simplest case, and many problems which have been tackled successfully in classical econometrics remain open here: Nothing has been said about the problems arising if the explanatory variables contain lagged endogeneous variables, for instance, which would destroy the independence of the disturbances in (4.12) and would require an appropriate revision of the procedure. But these and similar problems are, in a sense, technicalities which are beyond my compass here. It is sufficient that the issue of developing appropriate estimation techniques for time-varying models be made clear.

The nature of economic laws as described in Chaps. 2 and 3 suggest, however, two more general econometric consequences.

The first observation is fairly straightforward: If coefficients are changing over time, there is no point in insisting in asymptotic properties (like consistency) of the estimators for coefficients.

The second observation relates to the estimation of simultaneous equations systems. It is well-known that a proper estimation of a system of interdependent equations leads to different estimates for the coefficients of a certain equation than those obtained by estimating the equation in isolation. This is the "simultaneous equations bias". Similarly, the coefficients estimated for a given system might change if another equation is added to the system: Signs of the estimated coefficients might change, and even identifiability might be destroyed (Schönfeld (3)).

This is a severe problem since economic models are always partial models, and it is hence always possible to add another equation, quite irrespective of how large the system is.

Hence the simultaneous equations bias is not taken sufficiently seriously if it is used simply as an argument in favor of estimating systems of equations simultaneously rather than estimating single equations: Regardless of how large the system is, the bias will be present.

What seems to be required, then, is a determination of the reliability of the estimates of a given system *within* the system, but *without* adding further equations.

This seems to be possible, in principle, since the simultaneous equations bias can be attributed to correlations between explanatory variables and disturbances, and we might try to estimate those correlations directly within a given model, or argue that the simultaneous equations bias will not be too severe if our estimated residuals do not correlate too much with our explanatory variables[15]. In a pioneering study, Pauly offered a first successful attempt to cover some simple cases in classical regression. It remains open, however, whether the method proposed in Sect. 4.3 lends itself to similar possibilities.

15 See Goldberger pp. 288–294 for an interpretation of the simultaneous equations bias.

5. The Nature of Macroeconomic Laws

Willst du dich am Ganzen erquicken,
So mußt du das Ganze im Kleinsten erblicken.
Johann Wolfgang Goethe

5.1 On Reductionism

This chapter sets out to argue that there is nothing particularly disreputable in doing macroeconomic analysis even if no explicit microeconomic foundation is available. It will be argued, in fact, that macroeconomic laws might possess a certain degree of independence from their microeconomic underpinnings, and that analogies between microeconomic and macroeconomic laws might be misleading.

I object, hence, to the widespread "reductionist" view that microeconomics is more fundamental than macroeconomics. The reason given for that view is simply that the behaviour of aggregates is caused by individual decisions[1]. This is certainly true, in some sense, and nobody will dispute it, but this is not the problem.

The issue is not to assert some kind of irreconcilability between microeconomics and macroeconomics; an assertion which would, in fact, be incomprehensible to most (and to me). It is rather a possible difference in quality which we encounter, for instance, in thermodynamics when the *movement* of molecules on the micro level manifests itself as *heat* on the macro level[2]. Another example is the interaction of our cells which leads to conscious utility maximization. Incidentally our shopping behaviour is a macrophenomenon in this sense, and a strict reductionist ought to start from neurophysiology or the zoo of particles rather than from utility maximization for analysing it. But this does not seem to be a particularly sensible approach, and nobody seems to recommend it seriously. Rather it will be argued that certain stable relationships arise on the macro level of human conscience, perhaps for reasons of natural selection and social conditioning, which lend themselves to description by means of a stable utility function, rendering an explicit discussion of the underlying chaos redundant. An analogous

---

1 Cf. Hahn p. 36, for instance.
2 A similar example can be found in Becker.

argument pertaining to the relationship between macroeconomic and microeconomic laws will be developed in this chapter.

The main question to be dealt with is, of course, how macroeconomic laws emerge out of a given microeconomic structure. It will be argued that this can be analysed by a straightforward generalization of the isolating approach underlying the moving equilibrium method. This has various consequences regarding the stability of macroeconomic laws, their interdependence, and their shape. In all this, macroeconomic laws might differ from their microeconomic counterparts.

5.2 The Macroeconomic Method

5.2.1 Macroeconomics as an Isolating Approach

Macroeconomics deals with relationships between economic aggregates such as national product, employment, inflation etc. without taking explicit recourse to the underlying microeconomic variables like the income and employment situation of the individual households, and the movement of the various prices, etc. The implicit assumption of this procedure is obviously that these aggregate relationships exist and are sufficiently stable. But what is involved in this assumption? One reading would be to say that it is not necessary to take microeconomic variables into explicit consideration because there are microeconomic influences which enforce a certain microeconomic structure for each particular macroeconomic state. This underlies Keynes' analysis when he writes "... we assume that a given volume of effective demand has a particular distribution of this demand between different products uniquely associated with it"[3]. Another way of putting it is to say that macroeconomics presupposes that structural effects are negligible.

In the language of Chap. 2 we can rephrase this argument by saying that macroeconomic relationships can be isolated from microeconomic events: We can take the microeconomic processes enforcing definite microeconomic structures for any given macroeconomic state as data of our analysis, and can fix them under a ceteris paribus clause. Seen in this way, macroeconomic analysis is simply a particular way of using the isolating approach.

3 Keynes (1) p. 43 n.

There is another reading of macroeconomics, of course, which considers macroeconomic models as hypothetical ones which aim at developing certain ideas and mechanisms which are to be incorporated into more elaborate microeconomic structures later on[4]. This view – viewing macroeconomics models as generated by hypothetical isolations, as parables – seems to pose no theoretical problem, and hence we will not be concerned with it in the following. The theoretical problem arises if we ask ourselves what might be the substantive content of macroeconomic models which lack explicit microeconomic underpinnings. The issue is substantive isolation.

5.2.2 The Aggregation Principle

This and the following two sections describe the general nature of the relationship between microeconomic and macroeconomic laws, providing a starting point for our later discussions. Consider a microeconomic system. Denote the microeconomic state by the real vector x and its change over time by $\dot{x}$. The microeconomic system (S) is identified with its law of motion[5]

$$\dot{x} = f(x), \quad x, \dot{x} \in \mathbb{R}^m. \tag{S}$$

This system (S) is assumed to describe the microeconomic process truly, it is the *true system.*

Consider now an *aggregation rule* (A) which associates to each microeconomic state x a certain vector X of aggregate variables. It is assumed that the *macro state* X involves less components than the *micro state* x:

$$X = \alpha(x), \quad x \in \mathbb{R}^m, \ X \in \mathbb{R}^n, \ n < m. \tag{A}$$

The change of the macro state over time, denoted by $\dot{X}$, is conceived as a function of the macro state alone, i.e.

$$\dot{X} = F(X), \quad X, \dot{X} \in \mathbb{R}^n. \tag{M}$$

This is the *macroeconomic system (M).*

4 Fisher p. 575, Schlicht (1) pp. 52 ff.

5 For the sake of simplicity I refrain from more general formalizations here and assume, furthermore, that all differentiability and regularity conditions implicitly required by the arguments which follow are satisfied.

Assume that all this is given – a microeconomic system (S), an aggregation rule (A), and a macroeconomic system (M). In which sense can we expect the macroeconomic system (M) to describe the movements of the true microeconomic system correctly? This is the SAM problem.

A correct description would require that the true movement of the micro state leads, upon aggregation, to the same movement of the macro state as the macro system, i.e.

$$F(\alpha(x)) = \alpha_x(x) \cdot f(x). \tag{5.1}$$

This is certainly not achievable, in general, for all $x \in \mathbb{R}^m$ [6]. We might restrict our argument, however, to the set where this condition is fulfilled.

$$\Sigma := \{x \mid F(\alpha(x)) = \alpha_x(x) \cdot f(x)\}. \tag{5.2}$$

This is the *aggregation set,* i.e. the set of microeconomic states where the aggregate system holds true exactly. If we can give arguments assuring that all solutions of the microeconomic model approach this aggregation set very quickly (implying this set to be nonempty), we are justified in restricting our attention to the macroeconomic model: We can be sure that the microeconomic state is within the aggregation set already and that the macroeconomic model will give a good description of the movement of the macroeconomic aggregates.

This can be summarized in the following *aggregation principle: A necessary condition for a macroeconomic model being substantively isolated from its microeconomic underpinnings is that the solutions to the microeconomic system approach the aggregation set sufficiently fast.*

It is to be noted here that the macro system is only "valid" on the macro image of the aggregation set, i.e. on $\alpha(\Sigma)$. Hence we have to restrict the domain of the macro model to $\alpha(\Sigma) \cap \mathbb{R}^n$.

It is to be noted here that the aggregation principle states a requirement which is to be satisfied if a given macro model is to be substantively isolated. This does not at all exclude that alternative macro models – even different macro models based on the same aggregation rule – satisfy this principle.

6 Take e.g. linear functions $f(x) = Ax$, $\alpha(x) = Bx$, $F(X) = CX$ with constant matrices of appropriate order. Condition (1) would imply $CB = BA$ which cannot be achieved, in general, and it is not possible in general either to find for given A and B a matrix C which satisfies this condition. (See Schlicht (6) however.)

Consider for instance the case of two macro models F^1 and F^2 with aggregation sets $\Sigma_1 \supset \Sigma_2$. If F^2 satisfies the aggregation principle, F^1 will satisfy the aggregation principle, too, since all states approaching Σ_2 sufficiently fast will approach Σ_1 a fortiori. We might argue, of course, that F^1 is a better model because the associated aggregation set is larger. This argument is the same as that in Sect. 3.3.4 where it was also argued that the moving equilibrium approximation of the Marshallian market is not necessarily the best method. But the appropriateness of a macro model is not simply a matter of the size of the aggregation set. Rather the model is to be chosen such that the solutions enter the aggregation set as early as possible and remain therein. This has been the reason for choosing in Sect. 3.3.2 the slowest trajectory A (Fig. 1 p. 35) as the aggregation set rather than some other trajectory such as B. (Note that A and B cannot be compared by size easily.) We shall come back to alternative aggregation procedures in Sect. 5.4.4.

5.2.3 Aggregation and Moving Equilibrium

Consider the microeconomic system

$$\dot{x} = f(x) \tag{5.3}$$

and an aggregation rule

$$X = \alpha(x). \tag{5.4}$$

Introduce now additional variables Y and an additional aggregation rule $\beta\colon \mathbb{R}^m \to \mathbb{R}^k$, $k \geqq m - n$ such that there is a one-to-one correspondence between the vectors (X, Y) and x. Write

$$(X, Y) = \tilde{\alpha}(x) := (\alpha(x), \beta(x)) \tag{5.5}$$

and call it the augmented aggregation rule. Since $\tilde{\alpha}$ is assumed to be bijective, the true system (5.3) implies the equivalent system

$$\dot{X} = G(X, Y) \quad G := \alpha_x(\tilde{\alpha}^{-1}(X, Y)) \cdot f(\tilde{\alpha}^{-1}(X, Y)), \tag{5.6}$$

$$\dot{Y} = H(X, Y) \quad H := \beta_x(\tilde{\alpha}^{-1}(X, Y)) \cdot f(\tilde{\alpha}^{-1}(X, Y)). \tag{5.7}$$

If the movement of the auxiliary variables Y is sufficiently fast, the system (5.6), (5.7) can be analysed by the moving equilibrium technique along the lines indicated in Chap. 2. This is precisely the view of aggregation adopted here: Convergency towards the aggregation set $\Sigma := \{x \mid H(\alpha(x), \beta(x)) = 0\}$ is assumed to be sufficiently pronounced. Hence it turns out, in fact, that aggregative analysis is simply a variant of the isolating approach underlying the moving equilibrium method.

Seen the other way round, the moving equilibrium method is simply a special case in aggregation. The variables x (fast) and y (slow) are transformed by means of the aggregation rule $Y = \alpha(x, y)$ and the auxiliary variables are the fast variables x in the "direct" moving equilibrium case. The basic feature is still that a system of higher dimensions is described in lower dimensions. This, however, is the essence of aggregation.

5.2.4 The Macroeconomic Isolation Principle

The aggregation principle – that the solutions of the true system are almost always close to the aggregation set – states a necessary but by no means sufficient condition for a macroeconomic model to be substantively isolated from its microeconomic underpinnings. We have to consider, in addition, the problems of causal and temporal isolation. These are slightly more intricate.

Let us start with an example. Consider the statement: "If national income rises by 1%, consumption demand will increase ceterius paribus by 0.7%". The ceteris paribus clause refers here to a constancy of consumption behaviour, obviously, but also to the constancy of prices or to a definite reaction of consumer prices to a change in consumer demand, for instance, and to a definite distribution of the additional income among consumers. It is obvious that a given increase in national income might lead to quite different increases in consumption demand according to how it is distributed. (Only if all consumers have the same marginal propensity to consume will the demand increase remain unaffected by the distribution of income, but this is obviously a rather uninteresting case.)

Hence the ceteris paribus clause underlying our macroeconomic consumption function implies a definite association between the level of national income and its distribution. This would be purely hypothetical if we could not give reasons for this assumption. But this is not very difficult: It is sufficient to postulate that the true system (S) produces a definite income distribution and hence a definite association between national income and consumption demand. This can be done without knowing the true system (S) in detail, but it implies that the true system remains constant: The ceteris paribus clause entails not only the constancy of individual consumption behaviour, but also the constancy of the true system (S), since this system as a whole determines the distribution of income. Microeconomic production functions fall under the ceteris paribus clause, for instance, because they affect factor demand, factor prices and, hence, the distribution of income.

Generalizing we can say that a ceteris paribus clause, if employed in macroeconomic analysis, fixes the underlying microeconomic system. If we aim for substantive isolation, we have to consider whether this is a good approximation. In order to do that, we have to consider the impact of *changes* in the microeconomic system on the macroeconomic laws.

Denote by $\mathscr{F}^m$ the set of continuously differentiable functions $\mathbb{R}^m \rightarrow \mathbb{R}^m$ which contain the possible candidates for microeconomic systems and denote by $P: \mathscr{F}^m \rightarrow \mathscr{F}^n$ an *aggregation procedure* which associates a macroeconomic system $F \in \mathscr{F}^n$ to each microeconomic system $f \in \mathscr{F}^m$

$$F = P(f), \quad f \in \mathscr{F}^m, \; F \in \mathscr{F}^n. \tag{P}$$

The aggregation procedure (P) replaces the macroeconomic system in the SAM problem: We have a SAP problem (microeconomic system (S), aggregation rule (A), and aggregation procedure (P)).

Changes in the microeconomic system over time (due to changes in tastes and technology, for instance) will induce changes in the macroeconomic system according to (P). The conclusions drawn from a macroeconomic analysis under the assumption of a fixed microeconomic system will remain valid in an approximate sense, however, if the microeconomic solutions approach the aggregation set quickly, and if the aggregation set actually does not move too quickly: The movement of the aggregation set, as induced by changes in the microeconomic system, ought to be slow compared to the speed with which the microeconomic solutions approach the aggregation set in order not to supersede or destroy the conclusions drawn from the macroeconomic model. This amounts to what can be termed *macroeconomic temporal isolation.* Furthermore, causal isolation of the underlying microeconomic model requires that its data remain sufficiently unaffected by changes in its variables. This translates, in the macroeconomic context, into the requirement that the macroeconomic data (i.e. the shape of the various macro functions) remain sufficiently unaffected by movements in the macro variables. This amounts to what can be termed *macroeconomic causal isolation.*

A case where this requirement is violated is given by Keynes when he writes:

> The fault of the classical theory lies, not in limiting its terrain by assuming constant income, but in failing to see that, if either of its own variables (namely the propensity to save and the schedule of marginal efficiency of capital) change, income must ceteris paribus change; so that its tool breaks in its hand and it doesn't know and can't tell us what will happen to the rate of interest, when either of its own variables changes (Keynes (2), p. 559).

Hence in order that a macroeconomic theory be substantively isolated it is to be required that

1. The aggregation principle is satisfied; i.e. that all solutions of the underlying microeconomic model tend to the aggregation set sufficiently fast.
2. The macroeconomic temporal isolation principle is satisfied; i.e. that the movement of the aggregation set caused by changes in the underlying microeconomic system is sufficiently slow.
3. The macroeconomic causal isolation principle is satisfied; i.e. that the macroeconomic data remain sufficiently unaffected by the macroeconomic variables.

If these requirements are satisfied we say that the *macroeconomic isolation principle* is satisfied.

This formulation might seem to be much too demanding since it might suffice to say that the conclusions drawn from the macroeconomic model ought to be correct, and all deviations not affecting these conclusions ought to be permitted. This argument appears misleading however, since it leaves scope for correct conclusions drawn from wrong premises, and we shall stick to the view that a theory can be wrong even though it gives correct predictions. Hence the above, somewhat more elaborate, characterization has been chosen.

5.3 Closed Aggregation and the Context Dependency of Economic Laws

5.3.1 Closed Versus Open Aggregation: A Comparison in Statics

The usual view of the aggregation problem differs substantially from what has been said here, and the following is intended to put this difference into perspective.

The view of aggregation developed here will be called *closed aggregation* because it refers to a complete system which might be partial but is closed in the sense that the system determines the behaviour of all its endogeneous variables. Much of the aggregation literature is, in contrast, concerned with the problem of *open aggregation,* which is concerned with aggregating a subset of microeconomic relationships. These constitute an open system, typically, since the remaining equations which close the microeconomic model are not taken into consideration.

The distinction between open and closed aggregation (although not under these particular labels) is present already at the roots of the formal aggregation literature: The classic papers by Klein (1946) and May (1947) elaborate on this very issue: Klein favors open aggregation, and May argues in favor of closed aggregation. The economics profession as a class has preferred the Klein position and seems to have discarded May's arguments[7].

This appears to me to have been an unsound choice, and I shall try to explain by contrasting these two views in static analysis. Such an analysis allows for a modern reformulation of May's view, which will be seen to be just a special case of the view of aggregation presented in the foregoing sections.

---

7 One can only guess why this has been so. Perhaps the main reason is that May's procedure is mathematically trivial in the static context and offers no particular problems, whereas Klein's program posed a lot of challenging mathematical questions which turned out only later on to be answerable in the negative (though May had anticipated these problems). One should not forget to mention the contributions by Dresch, Pu, and May (1) as important forerunners of May's (2) seminal work and as a starting point for Klein's (1) most influential contribution.

We turn our dynamic models (S) and (M) into static models by setting the time derivatives to zero. Hence

$$f(x) = 0, \quad x \in \mathbb{R}^m \tag{SE}$$

denotes our microeconomic equilibrium model, and

$$F(X) = 0, \quad X \in \mathbb{R}^n,\ n < m \tag{ME}$$

is the static macro model. The aggregation rule

$$X = \alpha(x) \tag{A}$$

remains.

Closed Aggregation. Consider the set of microeconomic equilibria

$$E_f := \{x \mid f(x) = 0\} \tag{5.8}$$

and the set of macroeconomic equilibria

$$E_F := \{X \mid F(x) = 0\}\,. \tag{5.9}$$

The aggregation problem is solved if the macro model and the aggregation rule are selected such that

$$G_F = \alpha(G_f) \tag{5.10}$$

i.e. the aggregation of microeconomic equilibria gives the set of macroeconomic equilibria[8]. In the terminology of the foregoing section, the set of microeconomic equilibria is required to be contained in the aggregation set.

Condition (5.10) is not, in fact, very demanding. If the macro equilibrium is unique, for instance, $\alpha(G_f)$ contains one point $\bar{X}$ only (the macro equilibrium), and there are many macro systems F leading to the same equilibrium e.g.

$$F^i(X) = \varrho(X^i, \bar{X}^i), \quad i = 1, 2, \ldots, n \tag{5.11}$$

---

8 $\alpha(G_f) := \{\alpha(x) \mid x \in G_f\}$.

where the index i refers to the i-th macro variable and ϱ denotes the Euclidean distance. For the case that $\alpha(G_f)$ contains multiple equilibria, similar constructions can be made[9].

Hence the static problem of closed aggregation is essentially trivial, and many alternative macroeconomic systems $F \in \mathscr{F}^n$ will exist which have the same equilibrium set[10]. There thus remains much freedom concerning the appropriate choice of a macroeconomic system, and additional requirements could be introduced. Our foregoing discussion suggests that the macroeconomic model ought to be selected in such a way that its dynamic behaviour approximates the true microeconomic dynamics as closely as possible[11].

Open aggregation. Open aggregation is concerned with a different problem, which can best be illustrated and criticized by means of an example.

Consider n microeconomic consumption functions $c_i = \Psi(y_i)$, $i = 1, 2, \ldots, n$, which give the consumption expenditure c_i of individual i as determined by its income y_i. The aggregation problem is to find a macroeconomic consumption function $C = \Psi(Y)$ relating aggregate consumption $C = \Sigma c_i$ to aggregate income $Y := \Sigma y_i$. Hence the microeconomic variables are $c = (c_1, \ldots, c_n)$ and $y = (y_1, \ldots, y_n)$ and the aggregation rule is

$$\begin{pmatrix} C \\ Y \end{pmatrix} = \begin{pmatrix} 1, 1, \ldots, 1, & 0, \ldots, 0 \\ 0, 0, \ldots, 0, & 1, \ldots, 1 \end{pmatrix} \begin{pmatrix} c \\ y \end{pmatrix}. \tag{5.12}$$

The microeconomic system is

$$c - \psi(y) = 0, \quad \psi: \mathbb{R}^n_+ \to \mathbb{R}^n_+ \tag{5.13}$$

---

9 E.g. $F^1(X) = \varrho[X, \varrho(G_f)]$
$F^i(X) = 0$ for $i = 2, \ldots, n$.

This is not necessarily continuously differentiable, but might serve to illustrate the point. [Note that $\alpha(G_f)$ is closed if f and α are continuous.]

10 If M is any nonsingular $n \times n$-matrix and F is a macroeconomic system satisfying (5.10), the system $F' := MF$ will do the same, for instance.

11 As an additional advantage we might use Samuelson's "correspondence principle" if the macroeconomic model is chosen according to this criterion (cf. Samuelson, Chap. 9). An alternative point of view can be found in Klein (1) p. 93 who argues that the macroeconomic system ought to be chosen in such a way that it is analoguous to the underlying microeconomic system – he requires, for instance, that a marginal productivity theory ought to hold true for the macroeconomic system. If the macroeconomic system is selected in this way, however the use of the correspondence principle might be misleading. Still another way to select a macroeconomic model is to aim at reducing context dependency (see Sect. 5.4.4).

and the required macroeconomic system is

$$C - \Psi(Y) = 0, \quad \Psi: \mathbb{R}_+ \to \mathbb{R}_+. \tag{5.14}$$

Hence the problem is to find a function Ψ such that

$$\Sigma \Psi_i(y_i) = \Psi(\Sigma y_i) \tag{5.15}$$

for all $y \in \mathbb{R}^n_+$. Obviously this is possible if and only if the microeconomic consumption functions have the following special form

$$\Psi_i(y_i) = a_i + b\, y_i \tag{5.16}$$

for all i and certain constants $(a_1 \ldots a_n)$ and b, i.e. if all consumers have the same marginal propensity to consume. Hence open aggregation requires severe restrictions on the microeconomic functions, and similar severe "aggregation conditions" arise quite generally by attempts to replace several micro functions by a single macro function. Hence the view of open aggregation implies microeconomic restrictions which render aggregation impossible in almost all relevant cases; there seems to be consensus about this point[12].

This difficulty with open aggregation is not surprising if one looks at problem (5.12)–(5.15) from the point of view of closed aggregation. The microeconomic equilibrium set is

$$G_\psi := \{(c, y) \in \mathbb{R}^{2n}_t \mid c = \psi(y)\} \tag{5.17}$$

and the problem is simply to find a function $\Theta(C, Y)$ with the implied macroeconomic equilibrium set $\alpha(G_\psi)$[13]. As has been argued above (p. 72) this poses no difficulties[14].

This function would contain all information on the macroeconomic equilibria obtainable from the microeconomic system (5.13), and nothing more can be expected. On the other hand, this characterization of possible macroeconomic equilibria does not give very much information, in general, since the equilibrium set $\alpha(G_\psi)$ might be very large. If there exists, for instance, one household with a zero marginal propensity to consume and another household with a marginal propen-

12 See Nataf (2) p. 164 or Fisher p. 563, for instance. Klein's position is somewhat different and will be discussed in Sect. 5.5.2.

13 The aggregation rule α is given by (5.12).

14 $\Theta(C, Y) := \varrho((C, Y), \alpha(G_\psi))$ would do, for instance.

sity to consume of unity, the macroeconomic equilibrium set would be the entire area below the 45° line in the $Y-C$ diagram. Closed aggregation requires, therefore, a functional relationship $C=\Psi(Y)$ between the macroeconomic variables. The crucial point is, however, that such a functional relationship cannot be deduced from the microeconomic system (5.13) under reasonable conditions. It could be obtained by restricting the set of admissible income distributions. But this is precisely the point of view of closed aggregation: We take (5.33) as embedded into a larger microeconomic system which determines certain possible income distributions. This context restricts the set of macroeconomic equilibria and gives rise to meaningful macroeconomic relations[15]. Demanding (5.15) to hold true for *all* income distributions $y \in \mathbb{R}^n_+$ renders the aggregation problem sterile. This demand is excessive, however, since all we need to require is that (5.15) hold true for those income distributions which actually occur rather than for all income distributions $y \in \mathbb{R}^n_+$[16]. This amounts to discarding open aggregation and adopting the point of view of closed aggregation.

A more general formulation of the preceding straightforward: Take one group of equations of the microeconomic system (SE), say the first k equations, write the system in partitioned form as

$$\begin{aligned} &f^1(x)=0, \quad f^1\colon \mathbb{R}^m \to \mathbb{R}^k, \quad 1 \leqq k < m \\ &f^2(x)=0, \quad f^2\colon \mathbb{R}^m \to \mathbb{R}^{m-k} \end{aligned} \tag{5.18}$$

and partition the macro system similarly:

$$\begin{aligned} &F^1(X)=0, \quad F^1\colon \mathbb{R}^n \to \mathbb{R}^h, \quad 1 \leqq h < k \\ &F^2(X)=0, \quad F^2\colon \mathbb{R}^n \to \mathbb{R}^{n-h}, \quad h < n\,. \end{aligned} \tag{5.19}$$

The aggregation rule remains

$$X=\alpha(x)\,. \tag{A}$$

The partial macro systems F^1, F^2 are taken to subsume the partial micro systems f^1 and f^2, respectively.

15 This point of view is underlying the well-known Hicks-Leontief aggregation which presupposes constant relative prices within the aggregates in order to fix their structure (see Leontief (1), Hicks pp. 33 ff., Samuelson pp. 143 ff.).

16 Pu has stressed this point in detail in his criticism of Klein's views on aggregation.

Open aggregation is concerned with the partial systems

$$f^1(x) = 0 \tag{5.20}$$

$$F^1(X) = 0 \tag{5.21}$$

alone while "forgetting" about the remaining systems f^2 and F^2. The corresponding equilibrium sets are

$$E_{f^1} := \{x \in \mathbb{R}^m | f^1(x) = 0\} \tag{5.22}$$

$$E^o_{F^1} := \{X \in \mathbb{R}^n | F^1(X) = 0\}. \tag{5.23}$$

(The superscript o refers to open aggregation.) If the construction of F^1 uses only the information contained in f^1, nothing more than

$$E^o_{F^1} = \alpha(E_{f^1}) \tag{5.24}$$

can be obtained. Since $E^o_{F^1}$ might be very large, this will often be of limited use only.

From the point of view of closed aggregation, the remaining parts of the systems are taken explicitly into account. Define the equilibrium set belonging to f^2

$$E_{f^2} := \{x \in \mathbb{R}^m | f^2(x) = 0\}. \tag{5.25}$$

Since only equilibria in this set will be realized, we can replace (5.24) by the condition

$$E_{F^1} = \alpha(G_{f^1} \cap G_{f^2}) \tag{5.26}$$

which implies a smaller equilibrium set and hence a more informative partial macro system.

The aggregation condition (5.10) reads now

$$E_F := E_{F^1} \cap E_{F^2} = \alpha(E_{f^1} \cap E_{f^2}) \tag{5.27}$$

with

$$E_{F^2} := \{X \in \mathbb{R}^n | F^2(X) = 0\} \tag{5.28}$$

in the case of closed aggregation.

Compare this with the corresponding conditions for open aggregation

$$E^o_{F^1} = \alpha(E_{f^1}) \tag{5.29}$$

$$E^o_{F^2} = \alpha(E_{f^2}) \tag{5.30}$$

and look at the resulting macroeconomic equilibrium set

$$E_F^o := E_{F^1}^o \cap E_{F^2}^o. \tag{5.31}$$

This leads to the following theorem.

Theorem

1. The set of macroeconomic equilibria E_F obtained under closed aggregation is a subset of the macroeconomic equilibrium set E_F^o obtained under open aggregation, i.e.

$$E_F \subseteq E_F^o. \tag{5.32}$$

2. On the other hand, open aggregation might lead to macroeconomic equilibria with no corresponding microeconomic equilibrium, i.e.

$$E_F \neq E_F^o \tag{5.33}$$

might occur.

Proof. Relation (5.32) will be proved first. For any $X \in E_F$ there exists $x \in G_{f^1} \cap G_{f^2}$ such that $X = \alpha(x)$. Hence $X \in E_{F^1}^o$ and $X \in E_{F^2}^o$. That (5.33) might come to be can be shown by means of an example. Take $m = 4$, $n = 2$ and the microeconomic system

$$f^1(x) = \begin{pmatrix} x_1 \\ x_2 \end{pmatrix} \tag{5.34}$$

$$f^2(x) = \begin{pmatrix} x_3 - x_1 \\ x_4 - x_2 \end{pmatrix}. \tag{5.35}$$

This system has the unique equilibrium $x = 0$. Take the aggregation rule

$$X^1 = x_1 + x_2, \quad X^2 = x_3 + x_4. \tag{5.36}$$

Hence the macroeconomic equilibrium set obtained under closed aggregation contains only one point:

$$E_F = \{(0, 0)\}. \tag{5.37}$$

On the other hand, we have the equilibrium sets

$$E_{f^1} = \{x \in \mathbb{R}^4 \mid x_1 = 0, x_2 = 0\} \tag{5.38}$$

$$E_{f^2} = \{x \in \mathbb{R}^4 \mid x_1 = x_3, x_2 = x_4\} \tag{5.39}$$

implying

$$E^o_{F^1} = \{X \in \mathbb{R}^2 \mid X_1 = 0\} \tag{5.40}$$

$$E^o_{F^2} = \mathbb{R}^2. \tag{5.41}$$

Hence the macroeconomic equilibrium set obtained under open aggregation is

$$E^o_F = \{X \in \mathbb{R}^2 \mid X_1 = 0\} \tag{5.42}$$

which contains many points with $x_2 \neq 0$. These cannot be microeconomic equilibria, Q.E.D.

The basic criticism of open aggregation is contained in part 2 of the Theorem: Open aggregation might lead to wrong results.

5.3.2 The Context Dependency of Economic Relations

In comparing closed and open aggregation it turns out that under closed aggregation the macro relations F^1 depend on the whole microeconomic system, i.e. not only on f^1 but also on f^2. Furthermore, the partial macro system F^1 cannot be chosen independently of the remaining macro part F^2 if we require the aggregation condition (5.27) to hold true – as we should do. This feature – that a macro relation depends on the whole underlying micro system as well as on the remaining part of the macro system – can be referred to as *context dependency*.

This context dependency of macro relations is to be expected under closed aggregation whereas it is absent, by construction, in open aggregation. This problem was discussed at the very beginnings of the aggregation debate: Klein tried to avoid it, which led him to open aggregation; May, on the other hand, argued that we have to put up with it since it is impossible to derive meaningful macro relations, in general, if we insist upon context independency[17]. I have argued in the foregoing in favour of May's view, and the utterly restrictive results of aggregation theory following Klein seem to witness the same[18].

Hence, when talking in the following about macroeconomic relations, we will adopt the closed view and take into account that the relations involved might be context dependent. This does not imply, however, that context dependency is warranted, and we might still feel inclined to minimize it, in sympathy with Klein. Context dependency will simply turn out to be unavoidable in many cases, and instead of pretending it is not there, we will try to reduce it[19].

5.4 On Qualitative Differences Between Microeconomic and Macroeconomic Laws

5.4.1 Three Differences

At this point of the discussion it might be appropriate to pause and dwell a little on the qualitative relationship between microeconomic and macroeconomic laws from another point of view: Macroeconomic laws will not, in general, be simple analogues to the corresponding microeconomic relations. Their shape might be quite different, and simple analogies between microeconomic and macroeconomic relations (so often encountered in political arguments) might be quite misleading.

---

17 Klein (2) p. 309; May (2) p. 63. By the way, Keynes referred to context dependency when writing: "... an increase in effective demand due to an increased marginal propensity to consume might find itself faced by a different aggregate supply function from that which would face an equal increase in demand due to an increased inducement to invest" (Keynes (1) p. 43 n).

18 See Nataf (1) and Leontief (2), for instance.

19 By the way, the question of context dependency has some philosophical implications regarding the juxtaposition of the doctrines of "internal" and "external" relations in Russell, pp. 42–50. Internal relations are context dependent, external relations are not. If we argue – as in Sects. 5.5.1 and 5.5.2 below – that all relations can be viewed as macro because any system can be thought of as being produced by an underlying micro system, and if aggregation is bound to be closed, all relations ought to be expected to be context dependent, in principle. This contradicts Russell's view, which he considers as being of nearly logical necessity, that relations ought to be conceived of as external, and it vindicates the old dialectical view that each part of a system "contains" the whole of the system.

Even if a particular macroeconomic relation is simply a certain average of some microeconomic relations, it might look different because averaging involves some kind of *smoothing*. Furthermore, averaging might lead to an *elimination* of certain variables which are of microeconomic importance but then cease to be of macroeconomic importance. A similar elimination effect might occur in other ways, as we shall see later on. Furthermore, closed aggregation might lead to macroeconomic laws which have no qualitative counterpart in any of the underlying microeconomic relations, they might appear quite different. This is due to what might be denoted the *system effect*.

These three effects – the smoothing effect, the elimination effect, and the system effect – are closely interlinked, of course. For expositional reasons we will proceed to discuss them separately, however, offering brief characterizations and a few examples. These might serve to further illustrate the view on aggregation expounded here.

5.4.2 The Smoothing Effect

1. *General Characterization.* Putting microeconomic relations together in order to obtain a macroeconomic relation might smooth out microeconomic discontinuities. Consider the following examples.
2. *Microeconomic Fixed Proportions and Macroeconomic Variable Proportions*[20]. Take a certain number of firms. Each of them has a production function

$$y = \min\{a \cdot k, b \cdot n\} \tag{5.43}$$

relating output y to the inputs of capital k and labor n. The productivities a and b are distributed among firms according to the "generalized Pareto distribution"

$$f(a, b) = A a^{\alpha} b^{\beta}, \quad \alpha, \beta, A > 0\,. \tag{5.44}$$

20 This idea is from Houthakker; see also Johansen. Allen (2) pp. 400 ff. gives a similar effect arising in a vintage growth model with fixed proportions through technical progress. See also Solow/Tobin/v. Weizsäcker/Yaari. The "repairman problem" as analysed in Arrow/Levhari/Sheshinsky leads from micro substitution to macroeconomic fixed proportions. Hence we might sometimes find a kind of "sharpening effect", too.

As Houthakker has shown, the maximum aggregate output $Y=\int y$ obtainable from aggregate factor inputs $K=\int k$ and $N=\int n$ is given by a Cobb-Douglas production function

$$Y=CK^{\gamma}N^{\delta}, \quad \gamma:=(\alpha+1)/(\alpha+\beta+3), \quad \delta:=(\beta+1)/(\alpha+\beta+3). \tag{5.45}$$

Hence microeconomic fixed proportions might lead to macroeconomic variable proportions, and this idea is quite independent of the specific example chosen.

3. *Smoothing Demand Functions.* Consider a market for a certain commodity. Denote by y the income of household i, denote by p the price of the commodity, and by x_i the quantity demanded by household i which can either be zero or one. Individual demand is given by

$$x_i=\begin{cases}1 & \text{if } \; p\leqq\varphi(y_i)\;\varphi'>0\\ 0 & \text{if } \; p>\varphi(y_i)\end{cases} \tag{5.46}$$

i.e. individual demand is discontinuous. Thc household buys either one unit of nothing. Assume income distribution to be given by a decreasing function $F(y)$ which denotes the number of households having a income above y. Hencc market demand $x=\Sigma x_i$ is given by

$$x=F(\varphi^{-1}(p)) \tag{5.47}$$

which can be taken to be approximately continuous and smoothly decreasing (Fig. 3)[21].

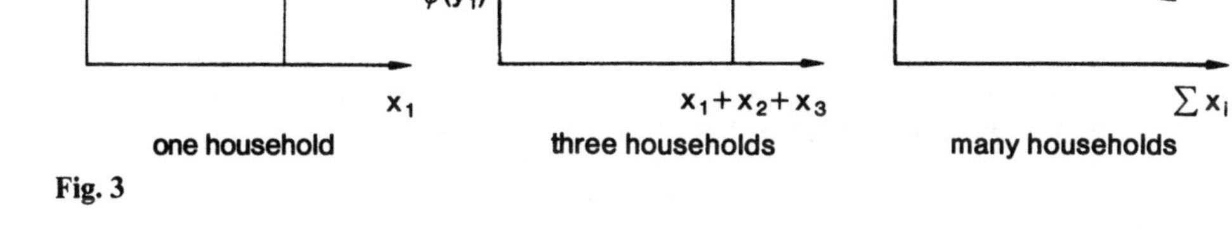

Fig. 3

21 Sondermann analyses this aspect of the smoothing effect much more generally. See also Duesenberry pp. 105 ff.

4. *Convexification.* A similar smoothing effect occurs when production sets or preferences are not convex. This leads, for the individual demander or supplier, to discontinuous demand and supply functions. These discontinuities smooth out in the aggregate, however, if many similar agents are present. If for a given price p the supplies x and x' are equally profitable for the individual firms, for instance, aggregate supply can be any number $nx+(N-n)\,x'$ where N denotes the total number of firms involved and n is the number of firms producing x. Since n can vary between zero and N, almost all intermediate supplies between Nx and Nx' can be realized in the aggregate. Hence aggregate supply would look approximately the same if all individuals firms had horizontal continuous supply functions between x and x' for price p emerging from convex production sets[22]. Similar arguments can be used for consumer demand[23].

5. *Aggregate Supply in a Fixprice Model.* Consider an economy with prices fixed in the short run. As a matter convenience, scale all commodites such that one unit has a price of one. Denote x_m^s the maximum supply of firm m which can profitably be sold at the given prices. If demand x_m^d larger than x_m, the firm will supply only x_m, if demand falls short of x_m^s, the firm will satisfy this demand. Hence the supply of an individual firm can be depicted by a 45-degree diagram with a capacity constraint at x_m^s (Fig. 4).

Now consider aggregate demand D. Along with Keynes we assume that "a given volume of effective demand has a particular distribution of this demand between different products uniquely associated with it"[24]. Hence we can write

$$x_m^d = x_m^d(D), \quad \sum_m x_m^d(D) = D \tag{5.48}$$

and we can look at aggregate supply if D varies

$$x = \sum_m \min\{x_m^d(D), x_m^s\}. \tag{5.49}$$

22 See Farell. This convexification argument is only applicable, of course, if the number of firms is sufficiently large and nonconvexities are small as compared to the size of the market. Strictly increasing returns, e.g. would not lead to this effect.

23 Such arguments are studied extensively in general equilibrium theory (see Arrow/Hahn Chap. 7).

24 Keynes (1) p. 43 n. Keynes view is somewhat different, however, since he is not arguing in a fixprice model. On the Keynesian supply function, see for instance Schlicht (3).

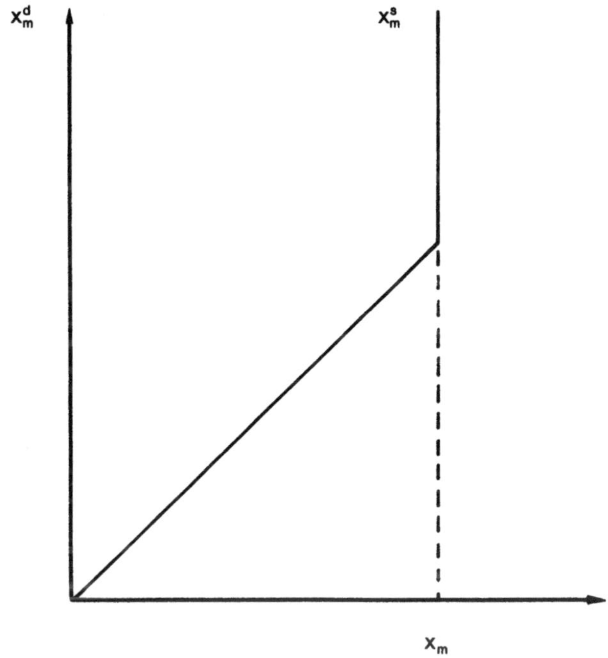

Fig. 4

If $x_m^d(D) \leqq x_m^s$ for all m, we will have $X = D$, leading to the familiar 45-degree diagram. If demand increases, however, some bottlenecks will occur, implying $x_m^d(D) > x_m^s$ for some firms. This leads to $X < D$ and the gap $D - X$ will increase if the number of firms running into their capacity limits increases with increasing aggregate demand. Hence the aggregate supply curve will look like the heavily drawn line in Fig. 5 rather than like the dashed curve. This implies, of course, that the distinction between "classical" and "Keynesian" unemployed in fixe-price models, which is based on the construction of Fig. 4 is somewhat blurred.

5.4.3 The Elimination Effect

1. *General Characterization.* If several microeconomic relations combine in a macroeconomic law, some microeconomic determinants might cease to be of influence on the macro level. Consider the following example:
2. *The Demand Functions of Duesenberry Households.* Consider a household with a utility function

$$u = u(x_1/a_1, x_2/a_2, \ldots, x_n/a_m) \tag{5.50}$$

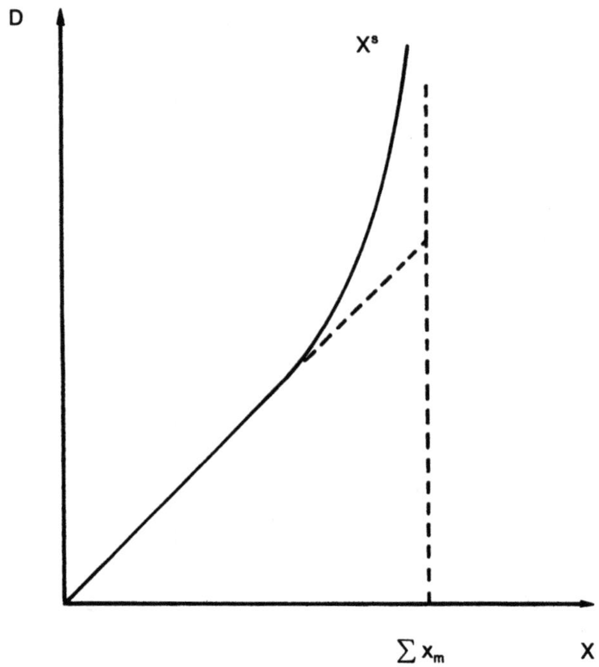

Fig. 5

where x_i denotes the quantity of commodity i consumed and a_i is the *aspiration level* with regard to commodity i. This household, which is called a Duesenberry household, receives utility from its consumption relative to its aspiration levels (see Duesenberry, p. 35).

Given income y and prices p_i, the household maximizes (5.50) under the budged constraint

$$\Sigma\, p_i\, x_i = y_i\,. \tag{5.51}$$

This leads to the necessary conditions

$$\frac{u_i}{u_j} = \frac{a_i\, p_i}{a_j\, p_j}\,. \tag{5.52}$$

Assume that these determine a unique maximum.

Consider now the formation of the aspiration levels a_i. These are considered as being determined by the consumption of the various commodities within the reference group of the household: a_i is a (possibly weighted) mean of the consumption of commodity i by the other households. Hence in equilibrium we can take $a_i = x_i$ for the typical household[25]. This transforms (5.52) into

$$\frac{u_i(1, 1, \ldots, 1)}{u_j(1, 1, \ldots, 1)} = \frac{x_i p_i}{x_j p_j}. \tag{5.53}$$

Together with (5.51) this implies

$$x_i = c_i \cdot \frac{y}{p_i} \tag{5.54}$$

with

$$c_i := \frac{u_i(1, 1, \ldots, 1)}{\sum_j u_j(1, \ldots, 1)}$$

constant.

Hence the aggregate demand system of the group of Duesenberry households forming a reference group is characterized by constant budget shares and can be rationalized by introducing a representative household with the utility function

$$u = x_1^{c_1} \cdot x_2^{c_2} \ldots x_n^{c_n} \tag{5.55}$$

which leads to the demand system (5.54) directly. Here aggregation leads to a simplification of the utility function through an elimination of the aspiration levels, and demand can be viewed as being generated by a representative household with a utility function (5.55) which is independent of aspiration levels.

5.4.4 The System Effect

1. *General Characterization.* Consider two microeconomic units which we want to describe by means of one aggregate relation. The behaviour of each unit is described by a function

$$v_i = f_i(d_i), \quad i = 1, 2. \tag{5.56}$$

where d_i denotes the data exogenous for the microeconomic agent, v_i denotes its response to these data. We are interested in a certain

25 On the notions of typical and representative agents see Sects. 1.2.5 and 1.2.6.

aggregation of individual actions

$$u = \varphi(v_1, v_2) \tag{5.57}$$

which could, of course, be described formally as determined by the data d_1 and d_2:

$$V = \Phi(d_1, d_2), \quad \Phi(d_1, d_2) := \varphi(f_1(d_1), f_2(d_2)). \tag{5.58}$$

It could be, however, that the data for individual 2 are partially determined by the actions of individual 1, and vice versa:

$$\begin{array}{c} v_1 = f_1(d_1) \\ \times \\ v_2 = f_2(d_2). \end{array}$$

If these mutual interdependencies are sufficiently strong, the relation (5.58) would violate the causal isolation principle, since the interdependency between variables and data cannot be neglected.

In order to take this into account, write these dependencies explicitly

$$d_1 = d_1(v_2, D) \tag{5.59}$$

$$d_2 = d_2(v_1, D) \tag{5.60}$$

where D denotes the data exogeneous for the system (5.58). Solving (5.58), (5.59), (5.60) yields

$$V = F(D) \tag{5.61}$$

which might describe a relationship between the variables V and D which appears quite different from either the relation between v_1 and D obtained from $v_1 = f_1(d_1(v_2, D))$ or the relation between v_2 and D obtained from $v_2 = f_2(d_2(v_1, D))$. The example of the Duesenberry households in Sect. 5.4.3 can be viewed as an illustration for the systems effect which leads, in that case, to an elimination of certain microeconomically important variables. The following example illustrates the systems effect from another angle and exemplifies the nature of the dynamic aggregation problems alluded to in Sect. 5.3.

2. *The Long-Run Savings Function in a Model of Distribution*[26]. Consider a neoclassical growth model: Labor productivity y is determined by the capital-labor ratio k according to the production function

$$y = f(k) \quad f(0) = 0,\ f' > 0,\ f'' < 0,\ f'(0) = \infty,\ f'(\infty) = 0. \qquad (5.62)$$

The wage rate w and the profit rate r are determined by the marginal productivities of labor and capital, respectively: There are n groups of individuals (or households), indexed by $i = 1, 2, \dots, n$, which are characterized by their savings ratio s_i $(0 < s_i < 1)$.

Denote by N_i the number of individuals belonging to group i and by k_i the capital holdings of group i. If all individuals receive the wage w, income of group i is[27]

$$y_i = w \cdot N_i + rK_i, \quad i = 1, 2, \dots, n. \qquad (5.63)$$

Savings lead to capital accumulation according to

$$\dot{K}_i = s_i \cdot Y_i \qquad (5.64)$$

Assume that all groups grow with the common growth rate g[28]

$$\dot{N}_i = gN_i, \quad i = 1, 2, \dots, n. \qquad (5.65)$$

Hence the relative group sizes

$$\alpha_i = N_i / \Sigma N_i, \quad i = 1, 2, \dots, n \qquad (5.66)$$

remain constant through time.

Define per capita wealth of group i as

$$k_i := K_i / N_i, \quad i = 1, 2, \dots, n \qquad (5.67)$$

and the capital-labor ratio as

$$k := \Sigma K_i / \Sigma N_i. \qquad (5.68)$$

---

26 The following is a variant of Stiglitz' distribution model.

27 Equation (5.63) presupposes only that the average wage and the average rate of profit is the same for all groups.

28 If N_i is measured in efficiency units, g is the sum of population growth and Harrod-neutral technical progress.

This implies together with (5.66) and (5.67) that the capital-labor ratio is the average of per capita wealth holdings

$$k = \Sigma\, \alpha_i k_i . \tag{5.69}$$

Furthermore (5.63)–(5.67) yield

$$\dot{k}_i = s_i (w + r \cdot k_i) - g k_i, \quad i = 1, 2, \ldots, n . \tag{5.70}$$

Together with (5.69) and (5.62), this defines the evolution of per capita wealth holdings, the capital-labor ratio, wages, and the rate of profit over time.

Our aggregation problem is the following: We take capital intensity k as our macro variable, i.e. we use the aggregation rule

$$\varkappa = \Sigma\, \alpha_i k_i \tag{5.71}$$

and aim for a Solow-type differential equation

$$\dot{\varkappa} = s(\cdot)\,(w(\varkappa) + r(\varkappa) \cdot \varkappa) - g\varkappa \tag{5.72}$$

such that the movement of k, as described by (5.62), (5.69), and (5.70) is approximated by the movement of $\varkappa$ as described by (5.62) and (5.72) while neglecting the structure of wealth distribution. How should we choose the aggregate savings ratio $s(\cdot)$, and what are the arguments we should insert into the brackets if we want to use aggregative variables only?

But before we consider this aggregation problem, we shall analyse the true model (5.70).

Consider Eq. (5.70). For any particular k and associated factor prices $w = w(k)$ and $r = r(k)$, we can define equilibrium per capita wealth holdings of group i as

$$k_i^* = k_i^*(k) := \frac{s_i w(k)}{g - s_i r(k)}, \quad i = 1, 2, \ldots, n . \tag{5.73}$$

If group i has $k_i = k_i^*$, k_i will remain stationary.

The equilibrium of system (70) can be defined, hence, by the conditions

$$k_i = k_i^*(k),\ \Sigma\, \alpha_i k_i = k . \tag{5.74}$$

Proposition. There exists a unique solution $(k_1, \ldots, k_n) \geqq 0$ of (5.74). It is a (locally) stable equilibrium of system (5.70).

Proof. Existence. Consider the function of k

$$k^*(k) = \Sigma\, \alpha_i k_i^*(k). \tag{5.75}$$

The requirement $k_i^* \geqq 0$ for all i, restricts the range of k to the set

$$\mathbb{K} := \{k \mid k > \underline{k}\}, \quad \underline{k} := \min_i r^{-1}(g/s_i) \tag{5.76}$$

For $k \in \mathbb{K}$, k^* is continuously differentiable. Since $k \downarrow \underline{k}$ implies $k^* \to \infty$ and $k \to \infty$ implies $(k^*/k) \to 0$, continuity implies existence.

Uniqueness. From (5.72) and (5.62) we find

$$\frac{dk_i^*}{dk} = \frac{f''}{w} k_i^*(k_i^* - k) \tag{5.77}$$

and hence

$$\frac{dk^*}{dk} = \frac{f''}{w} \Sigma\, \alpha_i k_i^*(k_i^* - k). \tag{5.78}$$

Evaluating this equation at equilibrium, and taking $f'' < 0$ into account leads to

$$\frac{dk^*}{dk} = \frac{f''}{w} \Sigma\, \alpha_i (k_i - k)^2 < 0 \quad \text{at} \quad k_i = k_i^* \quad \text{for all } i \tag{5.79}$$

which implies uniqueness.

Stability. Denote equilibrium by $(\bar{k}_1, \bar{k}_2, \ldots, \bar{k}_n)$ and write $\bar{k} = \Sigma\, \alpha_i \bar{k}_i$, $\bar{w} = w(\bar{k})$. Consider the Ljapunov function

$$V = \tfrac{1}{2} \Sigma\, (k_i - \bar{k}_i)^2 \frac{\bar{k}_i}{s_i \bar{w}} \tag{5.80}$$

Its time derivative is

$$\dot{V} = \Sigma\, (k_i - \bar{k}_i) \frac{\bar{k}_i}{s_i \bar{w}} \cdot \frac{s_i w}{k_i^*} (k_i^* - k_i) \tag{5.81}$$

with $w = w(k)$, $k_i^* = k_i^*(k)$.

The Taylor expansion around equilibrium yields the approximation

$$\dot{V} = \Sigma (k_i - \bar{k}_i) \left((\bar{k}_i - k_i) + \frac{f''(\bar{k})}{\bar{w}} \bar{k}_i (\bar{k}_i - \bar{k})(k - \bar{k}) \right). \tag{5.82}$$

This can be written in matrix notation. Define

$$z_i := k_i - \bar{k}_i, \quad b_i := \frac{f''(\bar{k})}{\bar{w}} \bar{k}_i (\bar{k}_i - \bar{k}) \tag{5.83}$$

and introduce the column vectors z, b, and α with components z_i, b_i, and α_i, respectively. Hence (5.82) is equivalent to

$$\dot{V} = z' (b\alpha' - I) z := z' A z. \tag{5.84}$$

The matrix $b\alpha'$ has roots $\mu_1 = \mu_2 = \mu_{n-1} = 0$ and

$$\mu_n = \alpha' b = \frac{f''}{w} \Sigma \alpha_i \bar{k}_i (\bar{k}_i - \bar{k}) = \frac{f''}{w} \Sigma \alpha_i (\bar{k}_i - \bar{k})^2 \leqq 0$$

(see Zurmühl p. 164). The matrix A has the roots $\lambda_i = \mu_i - 1$. Hence the range of A is $\mu_n - 1 \leqq z' A z / z' z \leqq -1$ for $z \neq 0$. This shows that V is decreasing around equilibrium, and this proves stability, Q.E.D.

Now consider the aggregation problem. As noted in Sect. 5.3.2 the aggregation procedure is not unique. Hence it might be appropriate to use our example for illustrating the nature of this nonuniqueness by discussing alternative aggregation procedures.

One way to determine the aggregate savings ratio in the macro equation

$$\dot{\varkappa} = s f(\varkappa) - g \varkappa \tag{5.85}$$

is to insert the aggregate savings ratio emerging in the true model in equilibrium. Denote it by $\bar{s}$:

$$\bar{s} := g \cdot \bar{k} / f(\bar{k}). \tag{5.86}$$

This is a constant determined by the data of the model. We have, in particular

$$\frac{\partial \bar{s}}{\partial s_i} = \frac{g w}{f^2} \cdot \frac{\alpha_i (w + r \bar{k}_i)}{1 - \partial k^* / \partial k} > 0 \tag{5.87}$$

$$\frac{\partial \bar{s}}{\partial g} = \frac{g w^2}{f^2} \cdot \Sigma \alpha_i \cdot s_i. \tag{5.88}$$

In other words, the aggregate savings ratio increases if individual savings ratios or the growth rate increase. Furthermore, $\bar{s}$ is dependent upon the shape of the production function. An additive shift β in the production function will yield, for instance,

$$\frac{\partial \bar{s}}{\partial \beta} = \frac{g}{f^2} \frac{\bar{k}}{1 - \partial k^* / \partial k} > 0 . \tag{5.89}$$

This equation illustrates some of the features of aggregative models noted in the text: The aggregate savings ratio is determined not only by the individual savings ratios, but by parameters such as the growth rate which are irrelevant microeconomically, and by the shape of the production function, which renders the macroeconomic savings ratio context dependent.

Now let us look for alternative aggregation procedures. Instead of using a constant macroeconomic savings ratio, we might consider it to be a function of our macro variable $\varkappa$:

$$s = s(\varkappa) . \tag{5.90}$$

As long as the resulting differential equation

$$\dot{\varkappa} = s(\varkappa) f(\varkappa) - g\varkappa \tag{5.91}$$

has the unique positive stable equilibrium $\bar{\varkappa} = \bar{k}$, this is a possible alternative. This implies, however, that $s(\bar{\varkappa}) = \bar{s}$. Hence the macroeconomic savings function will depend, directly or indirectly, on the growth rate and the shape of the production function, i.e. a qualitative difference between the micro and macro functions and context dependency cannot be avoided.

Furthermore, how should we select the function $s(\varkappa)$? "Optimal" aggregation would require choosing $s(\varkappa)$ such that the aggregation set covers the true solutions for most of the time (see the discussion in Sect. 5.2.2). Without going into mathematical details it might be sufficient here to point out that such a choice might lead to a canonical aggregation procedure (locally at least): Select that trajectory of the true system $(k_1(t), \ldots, k_n(t))$ which approaches equilibrium with the slowest speed, and look at the associated development of $k(t)$. If $k(t)$ is strictly monotone, take the savings ratio $s(k) := (\dot{k} + gk) f(k)$; if not, take $s(\varkappa) = \bar{s}$. This makes the aggregation set equal to slowest trajectory, and locally it implies that $s(\varkappa)$ is determined by equating the smallest characteristic root λ of the true system with the characteristic root of the aggregate system:

$$\frac{\partial \dot{k}}{\partial \varkappa} = s' f + s f' - g = \lambda . \tag{5.92}$$

Hence around equilibrium the "optimal" macro savings rate is given by

$$s(\varkappa) = \bar{s} + s'(\bar{\varkappa})(\varkappa - \bar{\varkappa}) = \bar{s} + (\lambda + g - \bar{s} f')(\varkappa - \bar{\varkappa}) \tag{5.93}$$

if the approach is monotone (Schlicht (6).)

Still another way to obtain an aggregate savings rate as a function of capital intensity is to aim not a comprehensive aggregation set, but rather at reduction of context dependency. We might aim, for instance, to write the savings function in such a way that the aggregate savings ratio depends on wages and profits, as determined by the production function – but so that the functional form of the savings ratio is independent of the production function. This can be done by treating system (5.70) in a way which resembles the moving equilibrium method very closely: Take the wage rate w and the profit rate r as fixed and look at the "temporary equilibrium" wealth holdings k_i^* resulting therefrom – as in (5.73) –

$$k_i^* = \frac{s_i w}{g - s_i r} . \tag{5.94}$$

For given factor prices, we obtain therefore an equilibrium capital intensity $k^* := \Sigma \alpha_i k_i^*$ and an associated savings rate

$$s^*(r) := \frac{\Sigma \alpha_i s_i (w + r k_i^*)}{w + r k^*} = \frac{\Sigma \alpha_i s_i/(g - s_i r)}{\Sigma \alpha_i/(g - s_i r)} . \tag{5.95}$$

As long as the individual stability conditions $g > s_i r$ are satisfied, (i.e. if $k \in \mathbb{K}$) this function is increasing in the rate of interest:

$$\frac{\partial s^*}{\partial r} > 0 . \tag{5.96}$$

Hence if we insert, in a second step, $r = r(k)$ according to (5.62), we obtain $s(\varkappa) = s^*(r(\varkappa))$, which can be inserted into the macro equation (5.91), and we again obtain the unique stable equilibrium $\bar{\varkappa} = \bar{k}$.

The savings ratio s^* is, however, independent of the shape of the production function: all influences entering from the production side are captured by the rate of interest. In this sense, context dependency is reduced by using s^*. But again we have a qualitative difference between micro and macro equations: The individual savings ratios are independent of the rate of interest whereas the latter influences, through its impact upon distribution and through the systems effect the aggregate savings ratio. Furthermore the aggregate savings ratio s^* depends upon the growth rate of the system (it is increasing with g), whereas the individual saving ratios are independent of that growth rate.

This example might illustrate, therefore, that there is no "correct" aggregation procedure, even for a given and well-defined problem: A model which minimizes context dependency will not necessarily offer a good dynamic approximation of the true movement (i.e. a comprehensive aggregation set), and vice versa. Our purpose will be decisive regarding the aggregation procedure actually chosen.

5.5 Hermeneutic Aggregation

5.5.1 The Status of the Aggregation Problem

The aggregation problem can be considered from two different perspectives: On the one hand we can ask how a given macroeconomic system can be described consistently in lower dimensions by a macroeconomic system. This is the *technical aggregation problem.*

On the other hand, however, we might ask quite generally how the nexus between microeconomic and macroeconomic models is to be conceived. This is the *hermeneutic aggregation problem.* It pertains in particular to those macroeconomic theories lacking an explicit microeconomic foundation, and this is practically the whole body of macroeconomics as pertaining to business cycles, foreign trade, distribution, money and employment. (Even if "microeconomic foundations" have been developed in some instances here, their elliptical or metaphorical character virtually excludes strict applications.) In these cases we might look for some justification for starting with macroeconomic models directly. What are the implicit premises of such a procedure, and what might be the meaning of those macroeconomic relations which have not been derived from explicit microeconomic foundations? This amounts to considering the aggregation problem from a hermeneutic perspective.

The technical aggregation problem poses merely technical rather than fundamental problems, since a microeconomic theory is to be presupposed here. We can use the microeconomic theory if we are unable to produce an aggregative version.

The hermeneutic aggregation problem is, in contrast, of a more fundamental character, since it pertains to all economic theories, in principle: all economic theories can be viewed as macro in the sense that any relationship we start with can be viewed as being generated by some underlying processes, and this generation gives rise to the hermeneutic aggregation problem – we cannot dispose of it a priori.

5.5.2 Revised Reductionism

Although the fundamental nature of the hermeneutic aggregation problem might be conceded in principle, one could still adopt a reductionist position by maintaining that micro laws are more fundamental than macro laws since the former "cause" the latter. Hence laws pertaining to lower levels of aggregation have a higher scientific status than laws relating to macro surface phenomena. In the words of Menger, "economic phenomena are *results* of the uncountable number of individual economic endeavours .... Hence we have to interpret them theoretically as such" (Menger, p. 87, my translation).

This revised reductionism, as we might call it, tends to deny the usefulness of aggregate theories as long as they have not been derived by technical aggregation, since, as Menger puts it, these theories can only be fictitious, and this renders the hermeneutic aggregation problem irrelevant.

5.5.3 Macroeconomic Order and Microeconomic Chaos

The premises of revised reductionism sketched above, namely, that laws are more fundamental the lower the level of aggregation to which they pertain, and that micro laws "cause" macro laws (and not vice versa) seems to be unfounded.

In order to explain this, consider the SAP problem discussed in Sect. 5.2.4: We have a set of microeconomic systems $\mathcal{F}^m$; an aggregation rule $\alpha: \mathbb{R}^m \to \mathbb{R}^n$; a set of macroeconomic systems $\mathscr{F}^n$; and an aggregation procedure P: $\mathscr{F}^m \to \mathscr{F}^n$ which associates a macroeconomic model $F \in \mathscr{F}^n$ to any given microeconomic model $f \in \mathscr{F}^m$.

Obviously microeconomic and macroeconomic models coexist. Neither the aggregation rule nor the aggregation procedure are conceived as having a causal claim[29]. They simply link two vistas of the same process, referring to two levels of aggregation selected from a spectrum of conceivable descriptions.

Furthermore, macro theories are *more general* than micro theories in the following sense: Typically the aggregation procedure will not be bijective, since different micro models might lead to the same macro model. Assume that all micro models out of a certain class C lead to the same macro model

$$P(f) = P(f') \quad \text{for all} \quad f, f' \in C \subset \mathscr{F}^m. \tag{5.97}$$

This macro model is more general than any micro model since it refers to the whole class C of micro models. It is, in the sense of Hayek, a *pattern* which is common to the whole class of micro models C. If there is a true micro model f which is stable, i.e. which remains unchanging over time, the associated macro model is stable. If the micro model changes within C, however, this microeconomic instability does not carry over to the macro model. Hence the macro model is not only more general, but more stable than the underlying micro model. Greater overall stability is, however, important in economics since it is often quite unreasonable to postulate stability throughout an economic system. We are bound to argue in a framework of ceteris paribus clauses, using the isolating approach, and are trying to understand economic regularities without spelling out the underlying psychological and physical laws in detail. Hence if we are interested in stable traits of economic processes, we have to select an appropriate level of aggregation, and there is no a priori presumption that lower levels of aggregation will exhibit more stability.

The mathematical chaos theory illuminates a related point. Consider a differential equation $\dot{x} = f(x)$ and consider its solutions $x(t, x_0)$ for alternative initial values. These solutions might turn out to be sensitive to inital values: The trajectories $x(t, x_0)$ and $x(t, x_0 + \varepsilon)$

29 I hope this statement is clear in spite of the unsolved ambiguity of the notion of causality: Philosophically speaking we do not know what "causality" means. See the review by Stegmüller, Chaps. V and VII.

diverge for arbitrarily small disturbances ε in the initial values. Such movements are termed chaotic:

> Small changes in initial conditions lead to large and unpredictable changes in the long-run evolution of the system .... This means that computations of long-run behaviour will be seriously affected by small errors. Furthermore, when a trajectory winds close to its initial point in phase space, sensitivity to initial conditions implies that its future evolution will, in general, be completely different from its past; in that sense, no long-range pattern can be discerned. An experimentalist who tries to describe the detailed behaviour of such a system will look in vain for reproducible trajectories, for unavoidable errors as well as noise during the experiment will conspire to produce a different pattern during each run. Both experimentally and computationally such a system would be described as chaotic or irregular[30].

Hence even if it is possibe to identify such a chaotic system exactly, which seems to be almost impossible, we could not use it for predictions since computational rounding errors are unavoidable.

Chaotic systems might be ergodic, however, in the sense that there exists a nontrivial function g of x such that the time average

$$\lim_{T \to \infty} \frac{1}{T} \int_0^T g(x(t, x_0))\, dt \tag{5.98}$$

is a well-defined vector of constants independent of initial conditions for almost all initial values. This ergodicity seems to be typical for dynamical systems, chaotic systems included[31]. Hence the chaotic micro system $\dot{x} = f(x)$ might give rise to a certain regularity in the macro variables defined as time averages of $g(x)$: Micro chaos and macro order might coexist.

Hence it might occur that stable structures occur only on a macro level: The underlying micro system might be unstable over time, or it might be chaotic. In these cases it seems to be futile to insist upon reductionism: We will not be able to detect the micro system; and we will not be interested in those chaotic movements as long as our aim is to describe regularities, and these occur on the macro level only. Hayek puts it as follows: "The fact is that in studies of complex phenomena

30 Yorke and Yorke p. 78. The best known and simplest example for a dynamic system leading to chaotic movements is the difference equation $x_{t+1} = a x_t (1 - x_t)$ with $3.57 < a < 4$. See Yorke/Yorke, Sect. 4.4 for a discussion of this equation. By the way, Day has recently introduced chaos theory into economics.

31 See Yorke and Yorke, p. 91. In a more restricted sense we might talk about ergodicity if the time average of x is independent of initial values, i.e. by choosing $g(x) = x$. The difference equation given in the preceding footnote displays ergodicity in this sense.

general patterns are all that is characteristic of those persistent wholes which are the main object of our interest, because a number of enduring structures will have this general pattern in common and nothing else" (Hayek (1), p. 345).

This does not imply, however, that microeconomic theories are to be discarded: They have their definite advantages, too. In particular they offer more detail and meaning in many cases. There seems to be no general a priori presupposition, therefore, that one approach is superior to the other. This will depend on the problem we are dealing with, and the questions we wish to answer.

Another aspect of the debate on reductionism is to be stressed: Very often it will be impossible to identify a micro system because of lack of data. This renders micro assumptions largely arbitrary, and macro assumptions covering a broader class of micro behaviour seem both more plausible – since they are less restrictive – and more easily testable. To take an example: It is well known from the work of Sonnenschein and Debreu that the general equilibrium model implies basically no restrictions on the system of excess demand functions other than the budget constraint. Hence no dynamic or comparative static results can be obtained without introducing additional restrictions. Restrictions on individual behaviour are, however, much more confining and more difficult to defend than aggregate assumptions summarizing the collective results of individual actions. If we want to make sure that those aggregate assumptions involve no contradictions, it suffices to present one single example, which can very often be provided by the assumption that all individuals are alike. This seems to be the main justification for using the concept of a typical agent[32].

Putting all these arguments together we conclude that macroeconomic relations are more general and more stable than the underlying micro relations, and are more accessible to the introduction of additional testable assumptions.

5.5.4 Structural Causality

The idea that invariant patterns emerge on the surface of a social system which appears rather irregular on the micro level has gained

32 A particularly important example of this type of argument is provided by Hildenbrand's "law of demand". On the concepts of typical and representative agents, see Sects. 1.2.5 and 1.2.6.

considerable impact in the social sciences. Sometimes it has been radicalized by maintining that micro relations are "determined" by macro relations. We shall henceforth speak of "structural causality" when referring to this holistic view [33]. Let us give some illustration.

1. *Selection by Competition.* Darwin's theory of natural selection explains the shape and behaviour of organisms by the principle of survival of the fittest. In analogy to that principle competition can be viewed as enforcing certain modes of behaviour by eliminating others (see Alchian). This argument gives the main reason for presupposing profit maximization, or growth maximization, or other competitively viable strategies on the side of the firms without recourse to psychology. In this sense, the competitive system enforces certain behavioural modes upon the agents through structural causation. Quite obviously competition itself is the outcome of individual behaviour, however, in the same sense as the Darwinian environments leading to natural selection can be viewed as being generated by the behaviour of organisms. It might nevertheless be fruitful to adopt the view of structural causation here if the outcome of competition is independent of individual behaviour for a wide spectrum of individual behavioural strategies, or if we can assume that the selective environment is sufficiently invariant with regard to the process under study that we are entitled to fix it under a ceteris paribus clause satisfying the isolation principle.

2. *Structural Causation in Marx.* In Marx, the idea of structural causation assumes a central position. He considers "capital" rather than "the capitalists" as one of the principal actors in the economic process. The capitalists are conceived as "character masks" which are forced by competition to act "in the interest of capital". He writes: "By and large, this is independent of benevolence or malice of the individual capitalist. Free competition enforces the immanent laws of capitalist production upon the individual capitalist from the outside" (Marx p. 268). His analysis attempts to explain the "universal laws of capitalistic accumulation" which manifest themselves in different countries in different specific historical appearances, and he draws a sharp distinction here: "The general and necessary tendencies of capital are to be distinguished from their outward appearances" (Marx p. 335). His theory aims at explaining those general laws which are thought to determine specific historical events. Only events and specific historical developments

33 Notwithstanding the philosophical problems associated with the notion of causality, see footnote p. 95 above.

serving those necessary tendencies will turn out to be relevant and override other possible influences[34]. Hence the macro patterns are thought of as determining, through structural causation, individual events.

3. *Structuralism.* The structuralist intent is quite similar: The aim is to explain a certain "structure" within which "events" realize. These events are thought of as being determined – although not uniquely determined – by the "structure".

In linguistics the structure-event juxtaposition appears as that of *"langue"* and *"parole"* (de Saussure pp. 9–25): language as a system of rules, determining possible sentences. In this sense the structure (*langue*) is prior to the event (*parole*) and can be analysed separately.

In an analoguous way, Lévi-Strauss has analysed marriage and kinship systems[35]: These systems are considered as sets of rules which determine possible marriages in the same way as the rules of grammar determine possible meaninful utterances. Various other ethnological problems – most notably the analysis of myths – have been tackled by following the paradigm of structural linguistics[36].

4. *Functionalism.* The idea of structural causation is of central importance here: Social facts are viewed as being determined by their function. They are not uniquely determined, however, since functionally equivalent solutions serve the same purpose.

Structural Causation: Dangers of the View. The idea of structural causation has proved quite fruitful in the social sciences, as the foregoing remarks might have suggested. A "structure" is thought to emerge from a certain system and can be analysed in isolation. Furthermore this structure is thought to be invariant with regard to a broad class of changes in the underlying system, and this asymmetry leads to a causal interpretation: The structure determines the events rather than vice versa.

---

34 Althusser calls this "surdétermination" and considers this idea as "Marx' great theoretical revolution" (see Althusser pp. 161–177).

35 Lévi-Strauss (1). For a good survey of Lévi-Strauss' views, see Oppitz.

36 Language is considered the social fact par excellence in Structuralism. Contrary to Marx, structures are hence viewed as not being replaceable by consent of free individuals (Marx p. 92). This is obvious with regard to language since there cannot be consent without communication, and the rules governing communication have to be presupposed in advance. Hence the basic social structures are viewed as working unconsciously in principle and can only be discovered in retrospect (see Oppitz pp. 62 ff.).

The danger of this view is to forget about the fact that a macro system can only exist if a corresponding micro system exists. Hence a macro system might be independent of the precise nature of the underlying micro system, but only to a certain extent, not absolutely, and the notion of structural causality is to be viewed always with caution: taking into account that it results from an isolating approach which is subject to critical ceteris paribus conditions.

Proponents of structuralist views sometimes do not pay sufficient attention to these limitations. If Marx writes in the above-mentioned passage: "Free competition enforces the immanent laws of capitalist production upon the individual capitalist from the outside" this could suggest that the "immanent laws of capitalist production" exist independently of "free competition". This reading of Marx seems to be patently incomprehensible, however, and many criticisms of Marxism oppose similar "historical laws" which determine the facts, rather than being determined by them.

The reproach of idealism is put forward against structuralism and functionalism in the same vein: "Structures" or "functional requirements" are presupposed as determining events, without taking into consideration that these structures or functional requirements emerge from those very events which they are supposed to determine[37]. This criticism seems to be correct as far as it goes, but if it is interpreted as implying that the notion of structural causation is misleading in principle, this seems to neglect a possible stability of macro structures with regard to micro changes. This asymmetry provides the basic justification for using the notion of structural causation fruitfully, and no a priori argument can exclude this possibility[38].

---

37 For a defense of idealistic structuralism, however, see Lévi-Strauss (2) pp. 19 ff.

38 A related problem of some importance, which can be touched only in passing here, is involved in the Marxian notion of "ultimate determination" ("*Determination in letzter Instanz*"), see Balibar pp. 212–222. How can we interpret the view that a complex interdependent system – a society for instance – is "ultimately determined" by some of its parts, such as the system of production? One answer is provided through the present approach. This can be illustrated best by reference to the Marshallian market model discussed in Sect. 3.3.1. Here the fast variable (the price) is determined by the slow variable, and the slow variable ultimately determines, in consequence, its own development.

5.6 Adequate Aggregation

Returning to the idea that micro chaos might lead to stable macro structures, it can be argued that the chaotic micro system describes the aggregate behaviour of an underlying well-structured micro system, and so forth[39]. Hence we can imagine various possible levels of aggregation exhibiting stable and comparatively simple structures, and in between these levels complexity. It appears sensible, then, to take those levels of aggregation distinguished by stability and simplicity as our starting point for further analysis.

But, of course, the questions we pose require certain levels of aggregation by themselves: Microeconomic questions require microeconomic theories although macro considerations might contribute to solve them. On the other hand, macroeconomic questions cannot typically be analyzed fruitfully in microeconomic terms (see Sect. 5.5.3). Insisting a priori on a particular level of aggregation does not therefore seem very sound. We have to select our theory with respect to our aim.

Hence we have to select a level of aggregation on which we have reasons to believe we can tackle our problem and such that the relations on which our analysis is built are sufficiently stable with regard to the processes we want to study. Conclusions are to be obtained and the isolation principle satisfied. But there is no a priori rule that one particular level of aggregation is to be preferred irrespective of the problem under discussion. We are bound to select an aggregation level which is appropriate to the problem under consideration.

39 Note that the chaotic systems discussed in Sect. 5.5.3 can be so viewed.

6. Epilogue. Economic Imagination

> **The economic needs the three great intellectual faculties, perception, imagination, and reason: and most of all he needs imagination, to put him on the track of those events which are remote or lie below the surface, and of those effects of visible causes which are remote or lie below the surface.**
>
> **Alfred Marshall**

Leaning back again, we might look at all that has been said from a broader perspective. Emphasizing that economic thinking proceeds by isolating the relevant processes and stressing the hermeneutic nature of the aggregation problem amount to envisaging economic processes as embedded in a broader context. Furthermore, in deciding whether these processes possess a sufficient degree of autonomy to make it possible to treat them in isolation, we must rely primarily on disciplined imagination.

In this sense, the task of the economist is similar to that of the poet:

> Lost in thoughts, he taps his knee with his wandlike pencil, and at the same instant a car (New York licence plate) passes along the road, a child bangs the screen door of a neighboring porch, an old man yawns in a misty Turkestan orchard, a granule of cindergray sand is rolled by the wind on Venus, a Docteur Jaques Hirsch in Grenoble puts on his reading glasses, and trillions of other such trifles occur – all forming an instantaneous and transparent organism of events, of which the poet (sitting in a lawn chair, at Ithaca, N.Y.) is the nucleus. (Nabokov, p. 161.)

The task of the economist is slightly different, however, in that he views the various marionettes as strung together by the price system and the flow of commodities but not so tightly as to exclude signalling among them, or whimsical reactions. Going beyond the ambition of his physiocratic ancestors, he is not only interested in exhibiting this picture, but is looking for its inner mechanics: those causal chains below the surface which set the picture in motion. This undertaking requires isolating specific phenomena and considering the interaction of aggregates without forgetting about the whole; it requires reason and imagination.

Furthermore, since the economist happens to be a naturalist, he will be utterly dissatisfied with merely inventing his figures – he aims at devising them in such a way that the bustle of the real world is mimicked by them (or that the real world mimics his picture). This requires perception as well as imagination and renders his task somewhat more difficult than that of the poet. But, as the late Charles Th. Hecktik (wriggling in a rocking chair in his crammed smokey study) used to babble, the economist is sometimes rewarded with the opportunity to observe incidents which had previously been suggested by his picture.

References

Adorno Th W (1) (1970) Zur Logik der Sozialwissenschaften. In: Adorno Th W, Aufsätze zur Gesellschaftstheorie und Methodologie. Suhrkamp, Frankfurt a. M.

Adorno Th W (2) (1966) Negative Dialektik. Suhrkamp, Frankfurt a. M.

Alchian A (1950) Uncertainty, Evolution, and Economic Theory. Journal of Political Economy 58:211–221

Allen R G D (1) (1965) Mathematical Economics, 2nd ed. Macmillan, London

Allen R G D (2) (1967) Macro-Economic Theory. Macmillan, London–Melbourne–Toronto

Athans M (1974) The Importance of Kalman Filtering Methods for Economic Systems. Annals of Economic and Social Measurement 3:24–49

Althusser L (1965) L'Object du Capital. In: Althusser L, Balibar É, Establet R (eds) Lire le Capital, Tôme II. Maspero, Paris

Arrow K J, Levhari D, Sheshinsky E (1972) Production Function for the Repairman Problem. Review of Economic Studies 39:24–49

Arrow K J, Hahn F (1971) General Competitive Analysis. Oliver and Boyd, London

Balibar E (1965) Sur les Concepts Fondamentaux du Materialisme Historique. In: Althusser L, Balibar É, Establet R (eds) Lire le Capital, Tôme II. Maspero, Paris

Becker G S (1962) Irrational Behavior and Economic Theory. Journal of Political Economy 70:1–13

Becker G S , Stigler G (1977) see Stigler/Becker

Bellman R (1960) Introduction to Matrix Analysis. Mc Graw Hill, New York–Toronto–London

Benassy J P (1975) Neo-Keynesian Disequilibrium Theory in a Monetary Economy. Review of Economic Studies 42:503–523

Bhatia N P, Szegö G P (1970) Stability Theory of Dynamical Systems. Springer, Berlin Heidelberg New York

Bhattacharyya M N (1976) An Evaluation of the Bonn Econometric Model via Time Series Analysis. Paper presented at the Econometric Society European Meeting in Helsinki (Otaniemi) August 1976, paper no. 73 C 2. University of Bonn, Institute of Econometrics, Bonn, Germany

Bliss C (1973) Comment. In: Mirrless J, Stern N H (eds) Models of Economic Growth. Macmillan, London, pp 51

Böhm-Bawerk E (1961) Positive Theorie des Kapitales. 4. Aufl. 1. Band. Fischer, Jena 1888; reprint Hain, Meisenheim

Clower R W (1969) The Keynesian Counter-Revolution: A Theoretical Appraisal. In: Clower R W (ed) Monetary Theory. Penguin, Harmondsworth, pp 270–297

Cooley T F, Prescott E (1973) An Adaptive Regression Model. International Economic Review 14:364–371

Darwin C R (1859) On the Origin of Species by Means of Natural Selection. Murray, London

Day R H (1982) Irregular Growth Cycles. American Economic Review 72:406–414

Debreu G (1974) Excess Demand Functions. Journal of Mathematical Economics 1:15–21

Dresch F W (1938) Index Numbers and the General Economic Equilibrium. Bulletin of the American Mathematical Society 44:134–141

Duesenberry J S (1949) Income, Saving, and the Theory of Consumer Behavior. Harvard University Press, Cambridge (Mass.)
Erwe F (1964) Gewöhnliche Differentialgleichungen. 2nd ed. Bibliographisches Institut, Mannheim
Eucken W (1947) Die Grundlagen der Nationalökonomie. 5th ed. Küpper, Bad Godesberg
Farrell M J (1957) The Convexity Assumption in the Theory of Competitive Markets. Journal of Political Economy 67:377–391
Fisher F M (1969) The Existence of Aggregate Production Functions. Econometrica 37:553–577
Friedman M (1953) The Methodology of Positive Economics. In: Friedman M (ed) Essays in Positive Economics. The University of Chicago Press, Chicago and London. pp 3–43
Galbraith J K (1972) Economics as a System of Belief. American Economic Review 62:1–10
Goethe J W (1949) Gedichte. In: E. Staiger (Hrsg) Band III. Manesse, Zürich
Goldberger A S (1964) Econometric Theory. Wiley, New York–London–Sydney
Hahn F H (1973) On the Notion of Equilibrium in Economics. University Press, Cambridge
Hahn F H, Arrow K J (1971) see Arrow/Hahn
Haken H (1978) Synergetics. 2nd ed. Springer, Berlin Heidelberg New York
Hamminga B (1982) Neoclassical Theory Structure and Theory Development: The Ohlin-Samuelson Programme in the Theory of International Trade. In: Stegmüller W, Balzer W, Spohn W (eds) Philosophy of Economics. Springer, Berlin Heidelberg New York. pp 1–15
Hatanaka H, Tanaka K (1980) On the Estimability of the Covariance Matrix in the Random Walk Representing the Time-Changing Parameters of the Regression Models. Paper presented at the Econometric Society World Congress Aix-en-Provence (France) 1980. Discussion paper no. 28, December 1980. Fakulty of Economics, Osaka University, Toyonaka, Osaka, Japan
Hayek F A (1) (1964) The Theory of Complex Phenomena. In: Bunge M (ed) The Critical Approach to Science and Philosophy. Free Press of Glencoe, London
Hayek F A (2) (1969) Freiburger Studien. Siebeck, Tübingen
Hayek F A (3) (1975) In Pretence of Knowledge. Nobel Memorial Lecture. Swedish Journal of Economics 77:433–442
Hicks J R (1946) Value and Capital. 2nd ed. Clarendon Press, Oxford
Hildenbrand W (1983) On the "Law and Demand". Econometrica 51:997–1019
Houthakker H S (1955/56) The Pareto Distribution and the Cobb-Douglas Production Function in Activity Analysis. Review of Economic Studies 23:27–31
Jevons W S (1976) The Theory of Political Economy, In: Collison Black R D (ed). Penguin, Harmondsworth
Johansen L (1972) Production Functions. North Holland, Amsterdam
Kaldor N (1972) The Irrelevance of Equilibrium Economics. Economic Journal 82:1237–1255
Kaldor N (1940) A Model of the Trade Cycle. Economic Journal 50:78–92
Kalman R E (1963) New Methods in Wiener Filtering Theory. In: Bogdanoff L, Kozin F (eds) Proceedings of the First Symposium on Engineering Applications of Random Functions Theory and Probability. Wiley, New York–London, pp 270–388
Keynes J M (1) (1936) The General Theory of Employment, Interest, and Money. Macmillan, London
Keynes J M (2) (1973) The General Theory and After, Part I: Preparation. The Collected Works of John Maynard Keynes, vol. XIII. Macmillan, London
Keynes J M (3) (1973) The General Theory and After, Part II: Defense and Development. The Collected Works of John Maynard Keynes, vol. XIV. Macmillan, London
Klein L R (1946) Macroeconomics and the Theory of Rational Behaviour. Econometrica 14:93–108
Klein L R (1946) Remarks on the Theory of Aggregation. Econometrica 14:303–312

Leontief W (1) (1936) Composite Commodities and the Problem of Index Numbers. Econometrica 4:39–59

Leontief W (2) (1947) Note on the Interrelation of Subsets of Independent Variables of a Continuous Function with Continuous First Derivatives. Bulletin of the American Mathematical Society 53:343 ff.

Levhari D (1972) see Arrow et al.

Lévi-Strauss C (1) (1949) Les Structures Elémentaires de la Parenté. Presses Universitaires de France, Paris

Lévi-Strauss C (2) (1964) Le Cru et le Cuit. Plon, Paris

Lexis W (1855) Grenznutzen. In: Handwörterbuch der Staatswissenschaften. 1. Supplement. Fischer, Jena, pp 422–432

Lindahl E (1939) Studies in the Theory of Money and Capital. George Allen and Unwin, London

Lotka A J (1956) Elements of Mathematical Biology. Dover, New York

Malinvaud E (1978) The Theory of Unemployment Reconsidered. Blackwell, Oxford

Markus L (1961) Structurally Stable Differential Systems. Annals of Mathematics 73:1–19

Marshall A (1949) Principles of Economics. 8th ed. Macmillan, London

Marx K (1970) Das Kapital, Erster Band. Dietz, Berlin

May K (1946) The Aggregation Problem for a One-Industry Model. Econometrica 14:285–298

May K (1947) Technological Change and Aggregation. Econometrica 15:51–63

Mc Closkey D M (1983) The Rhetoric of Economics. Journal of Economic Literature 21:481–517

Menger C (1883) Untersuchungen über die Methode der Socialwissenschaften und der politischen Ökonomie insbesondere. Mohr-Siebeck, Leipzig, reprint Tübingen

Mill J S (1965) On the Logic of the Moral Sciences. A System of Logic, Book VI. Magid H M (ed) Bobbs-Merrill, Indianapolis–New York–Kansas City

Nabokov V (1973) Speak, Memory. Pyramid, New York

Nataf A (1948) Sur la Possibilité du Construction de certains Macromodèles. Econometrica 16:232–244

Nataf A (1968) Aggregation. In: Sills D L (ed) International Encyclopedia of the Social Sciences, vol 1. Macmillan-Free Press, New York, pp 162–168

Oppitz M (1975) Notwendige Beziehungen. Suhrkamp, Frankfurt

Pagan A (1980) Some Identification and Estimation Results for Regression Models with Stochastically Varying Coefficients. Journal of Econometrics 13:341–363

Pauly R (1983) Offene versus vollständige Modellbildung in der Ökonometrie. Statistische Hefte 24:21–45

Prescott E, Cooley T F (1973) see Cooley/Prescott

Pu S S (1946) A Note on Macroeconomics. Econometrica 14:299–302

Russell B (1959) My Philosophical Development. George Allen and Unwin, London

Samuelson P A (1947) Foundations of Economic Analysis. Harvard University Press, Cambridge (Mass)

Saussure F de (1983) Course in General Linguistics, translated by R. Harris Duckwell, London

Schlicht E (1) (1976) Einführung in die Verteilungstheorie. Rowohlt, Reinbeck

Schlicht E (2) (1976) Die Methode der Gleichgewichtsbewegung als Approximationsverfahren. In: Helmstädter E (Hrsg) Neuere Entwicklungen in den Wirtschaftswissenschaften. Duncker und Humbolt, Berlin pp 293–305

Schlicht E (3) (1979) On Keynesian Supply. Discussion Paper No. 54. Fakultät für Wirtschaftswissenschaften Universität Bielefeld. 4800 Bielefeld 1, Federal Republic of Germany, January 1979

Schlicht E (4) (1982) Der Gleichgewichtsbegriff in der ökonomischen Analyse. Jahrbuch für Sozialwissenschaft 33:50–63

Schlicht E (5) (1977) Grundlagen der ökonomischen Analyse. Rowohlt, Reinbeck

Schlicht E (6) (1985) The Local Aggregation Problem. Discussion. Paper, Institut für Volkswirtschaftslehre, Technische Hochschule Darmstadt, Darmstadt
Schneider E (1964) Einführung in die Wirtschaftstheorie. 2. Teil, 9th ed. Siebeck, Tübingen
Schönfeld P (1) (1969) Methoden der Ökonometrie, Bd. 1. Vahlen, Berlin
Schönfeld P (2) (1971) Methoden der Ökonometrie, Bd. 2. Vahlen, Berlin
Schönfeld P (3) (1978) Probleme der ökonometrischen Modellbildung. In: Schober F, Plötzeneder H D (Hrsg) Ökonometrische Modelle und Systeme. Oldenbourg, München–Wien, pp 7–21
Schumpeter J A (1) (1954) History of Economic Analysis. George Allen and Unwin, London
Schumpeter J A (2) (1936) Review of the General Theory by Keynes. Journal of the American Statistical Association 31:792 ff.
Sebag L (1964) Marxisme et Structuralisme. Payot, Paris
Sheshinsky E (1972) see Arrow et al.
Solow R M, Tobin J, v. Weizsäcker C C, Yaari M (1968) Neoclassical Growth with Fixed Factor Proportins. Review of Economic Studies 35:117–124
Sondermann D (1975) Smoothing Demand by Aggregation. Journal of Mathematical Economics 2:201–223
Sonnenschein H (1972) Market Excess Demand Functions. Econometrica 40:549–563
Stegmüller W (1969) Probleme und Resultate der Wissenschaftstheorie und der analytischen Philosophie. Band 1: Wissenschaftliche Erklärung und Begründung. Springer, Berlin Heidelberg New York
Stigler G, Becker G S (1977) De Gustibus Non Est Disputandum. American Economic Review 67:76–90
Stiglitz J E (1969) Distribution of Income and Wealth Among Individuals. Econometrica 37:382–397
Szegö G P, Bhatia N P (1970) see Bhatia/Szegö
Tanaka K, Hatanaka H (1980) see Hatanaka/Tanaka
Thom R (1972) Stabilité Structurelle et Morphogenèse. Benjamin, New York
Tinbergen J (1940) On a Method of Statistical Business-Cycle Research. A Reply. Economic Journal 50:141–154
Tobin J (1968) see Solow et al.
Walras L (1954) Elements of Pure Economics. Translated by Jaffé. W. Allen and Unwin, London
v. Weizsäcker C C (1968) see Solow et al.
Williamson O (1975) Markets and Hierarchies. Free Press, New York
Yaari M (1968) see Solow et al.
Yorke J A, Yorke E D (1981) Chaotic Behavior and Fluid Dynamics. In: Swinney H L, Gollub J P (eds) Hydrodynamic Instabilities and the Transition to Turbulence. Springer, Berlin Heidelberg New York, pp 77–95
Zurmühl R (1964) Matrizen, 4th ed. Springer, Berlin Göttingen Heidelberg

Author Index

Subject Index

MIX
Papier aus verantwortungsvollen Quellen
Paper from responsible sources
FSC® C105338

If you have any concerns about our products,
you can contact us on
ProductSafety@springernature.com

In case Publisher is established outside the EU,
the EU authorized representative is:
Springer Nature Customer Service Center GmbH
Europaplatz 3, 69115 Heidelberg, Germany

Printed by Libri Plureos GmbH
in Hamburg, Germany